Climate Justice in India

Climate Justice in India brings together a collective of academics, activists, and artists to paint a collage of action-oriented visions for a climate just India. They offer historically and socially grounded perspectives on justice implications for Indian society, politics, and economics. This unique and agenda-setting volume informs researchers and readers interested in topics of just transition, energy democracy, intersectionality of access to drinking water, agroecology and women's land rights, national and state climate plans, urban policy, caste justice, and environmental and climate social movements in India. It synthesizes the historical, social, economic, and political roots of climate vulnerability in India and articulates a research and policy agenda for collective democratic deliberations and action.

This crossover volume will be of interest to academics, researchers, social activists, policymakers, politicians, and general readers looking for a comprehensive introduction to the unprecedented challenge of building a praxis of justice in a climate-changed world.

Prakash Kashwan is Associate Professor of Environmental Studies at Brandeis University, Waltham, Massachusetts. At the time of preparation of this volume, he was Associate Professor of Political Science at the University of Connecticut. He is the author of *Democracy in the Woods* (2017), editor of the journal *Environmental Politics*, and co-founder of the Climate Justice Network.

Climate Justice in India

Edited by
Prakash Kashwan

CAMBRIDGE
UNIVERSITY PRESS

CAMBRIDGE
UNIVERSITY PRESS

Shaftesbury Road, Cambridge CB2 8EA, United Kingdom

One Liberty Plaza, 20th Floor, New York, NY 10006, USA

477 Williamstown Road, Port Melbourne, vic 3207, Australia

314–321, 3rd Floor, Plot 3, Splendor Forum, Jasola District Centre, New Delhi – 110025, India

103 Penang Road, #05–06/07, Visioncrest Commercial, Singapore 238467

Cambridge University Press is part of the University of Cambridge.

It furthers the University's mission by disseminating knowledge in the pursuit of education, learning and research at the highest international levels of excellence.

www.cambridge.org

Information on this title: www.cambridge.org/9781009171915

First published 2022
Reprint 2023, 2024

Printed in India by Avantika Printers Pvt. Ltd.

A catalogue record for this publication is available from the British Library

Library of Congress Cataloging-in-Publication Data

Names: Kashwan, Prakash, editor.
Title: Climate justice in India / edited by Prakash Kashwan.
Description: Cambridge, United Kingdom ; New York, NY : Cambridge
   University Press, 2022. | Includes bibliographical references and index.
Identifiers: LCCN 2022020696 (print) | LCCN 2022020697 (ebook) | ISBN
   9781009171915 (hardback) | ISBN 9781009171908 (ebook)
Subjects: LCSH: Climatic changes--Government policy--India. | Climatic
   changes--Social aspects--Inida. | Environmental justice--India. | BISAC:
   LAW / Environmental
Classification: LCC QC903.2.I4 C57 2022 (print) | LCC QC903.2.I4 (ebook)
   | DDC 363.738/74560954--dc23/eng20220722
LC record available at https://lccn.loc.gov/2022020696
LC ebook record available at https://lccn.loc.gov/2022020697

ISBN 978-1-009-17191-5 Hardback

Cambridge University Press has no responsibility for the persistence or accuracy of URLs for external or third-party internet websites referred to in this publication, and does not guarantee that any content on such websites is, or will remain, accurate or appropriate.

Contents

Poems and Artworks

Tables

Figures

Abbreviations

ACCCRN	Asian Cities Climate Change Resilience Network
AKRSP [I]	Aga Khan Rural Support Programme (India)
AMRUT	Atal Mission on Rejuvenation and Urban Transformation
AP	Andhra Pradesh
ASUS	Ambedkar Slum Utthan Sangathan
BCCL	Bharat Coking Coal Limited
BJP	Bharatiya Janata Party
CBA	Chhattisgarh Bachao Andolan
CBDR	Common But Differentiated Responsibilities
CCL	Central Coalfields Limited
CDKN	Climate and Development Knowledge Network
CIL	Coal India Limited
CMM	Chhattisgarh Mukti Morcha
CoP	Conference of Parties
CRDPs	Climate-Resilient Development Pathways
CSE	Centre for Science and Environment
CSR	Corporate Social Responsibility
DGSM	Dasholi Gram Swarajya Mandal
EIA	Environment Impact Assessment
EIA/SIA	Environmental and Social Impact Assessments
ERR	Earthquake Rehabilitation and Reconstruction
FLCs	Forest Labour Co-operatives
FRA	Forest Rights Act

GDP	Gross Domestic Product
GEAG	Gorakhpur Environmental Action Group
GHGs	Greenhouse Gases
GIZ	German Agency for International Cooperation
GST	Goods and Services Tax
GW	Gigawatt
GWSSB	Gujarat Water Supply and Sewerage Board
HPERC	Himachal Pradesh Electricity Regulatory Commission
HRtWS	Human Right to Water and Sanitation
HSAA	Hindu Succession (Amendment) Act
IAS	Indian Administrative Service
ICJ	India Climate Justice
ICJF	Indian Climate Justice Forum
IMF	International Metalworkers Federation
INDCs	Intended Nationally Determined Contributions
ITUC	International Trade Union Confederation
IUCN	International Union for Conservation of Nature
JLG	Joint Liability Groups
JMACC	Jharkhand Mines Area Coordination Committee
JMM	Jharkhand Mukti Morcha
JNNURM	Jawaharlal Nehru National Urban Renewal Mission
JOHAR	Jharkhand Organization for Human Rights
KSSP	Kerala Sastra Sahitya Parishad
kWh	Kilowatt-hour
LARRA	Land Acquisition, Rehabilitation and Resettlement Act
LSGIs	Local Self Government Institutions
LVC	La Via Campesina
MAKAAM	Mahila Kisan Adhikaar Manch
MGNREGA	Mahatma Gandhi National Rural Employment Guarantee Act
MKSP	Mahila Kisan Sashaktikaran Pariyojana
MMDRA	Mines and Minerals (Development and Regulation) Act
MP	Member of Parliament
MP	Madhya Pradesh
MW	Megawatt
NABARD	National Bank for Agriculture and Rural Development
NAPCC	National Action Plan on Climate Change
NBA	Narmada Bachao Andolan
NCDHR	National Commission on Dalit Human Rights
NCRB	National Crime Records Bureau

NDCs	Nationally Determined Contributions
NESPON	North Eastern Society for the Preservation of Nature and Wildlife
NGO	Non-governmental Organization
NGT	National Green Tribunal
NHG	Neighbourhood Group
NRLM	National Rural Livelihoods Mission
NTPC	National Thermal Power Corporation
NTUI	New Trade Union Initiative
NVDP	Narmada Valley Development Project
OBC	Other Backward Classes
ODF	Open Defecation Free
PACS	Primary Agricultural Co-operative Societies
PESA	Panchayat (Extension to Scheduled Areas) Act
PESA	Panchayat (Extension to Scheduled Areas) Act
PIL	Public Interest Litigation
PMCCC	Prime Minister's Council on Climate Change
PSUs	Public Sector Undertakings
PV	Photovoltaics
RE	Renewable Energy
RECs	Rural Electric Co-operatives
REDD	Reducing Emissions from Deforestation and Forest Degradation
RESCO	Renewable Energy Service Company
RSS	Rashtriya Swayamsevak Sangh
RTI	Right to Information
SAPACC	South Asian People's Action on Climate Crisis
SAPCCs	State Action Plans on Climate Change
SC	Scheduled Castes
ST	Scheduled Tribes
SECL	South Eastern Coalfields Limited
SHG	Self-help Group
SIPB	State Investment Promotion Board
SSP	Sardar Sarovar Project
T-Zed	Towards Zero Carbon Development
TERI	The Energy and Resources Institute
TNRLM	Tamil Nadu Rural Livelihoods Mission
TNWC	Tamil Nadu Women's Collective
TUED	Trade Unions for Energy Democracy
UAPA	Unlawful Activities (Prevention) Act
UIDSSMT	Urban Infrastructure Development Scheme for Small and Medium Sized Towns

ULBs	Urban Local Bodies
UN	United Nations
UNDP	United Nations Development Programme
UNDRIP	United Nations Declaration on the Rights of Indigenous Peoples
UNDROP	United Nations Declaration on the Rights of Peasants and Other People
UNDROP	United Nations Declaration on the Rights of Peasants and Other People Working in Rural Areas
UNFCC	United Nations Framework Convention on Climate Change
UNHCR	UN High Commissioner for Refugees
UP	Uttar Pradesh
USAID	US Agency for International Development
VECs	village energy committees
WASMO	Water and Sanitation Management Organisation
WB	West Bengal
WCD	World Commission on Dams

Preface and Acknowledgements

My ongoing engagements with international and national debates on climate justice are a result of an intellectual journey over the past two decades that has brought me time and again to the complex intersections of environmental protection and social justice. Market-based solutions became the backbone of ostensible global responses to climate change at the United Nations Climate Change Conference held in Bali in December 2007. The sense of excitement among environmental economists then is difficult to describe from where we are today. However, to those of us who had spent time in the field, this euphoria was evidently and grossly misplaced. The journey that market-based solutions would have to take, from Bali to places like Bastar in Chhattisgarh, where they would be eventually implemented, is not paved with the freedom of choice that pro-market advocates like to celebrate.

Markets are designed to facilitate the accumulation of surplus in the hands of those who can channel it higher up in the 'food chain'. In most cases, the market ecosystem is essentially a centralizing force and does not work for the poor and marginalized. Unfortunately, this argument often falls through the cracks due to the lack of interdisciplinary work that is needed to produce knowledge that may help inform public debates on these complex questions. The market-based solutions institutionalized at the Bali climate conference, especially carbon offsets and carbon emissions trading, have proven to be colossal failures.

Perhaps even more embarrassingly, the advocates of market-based climate solutions lost the battle of ideologies to right-wing reactionary forces. Even in the supposedly knowledge-driven market economies of the Global North, ultra-conservatives have been successful in labelling neoliberal policies, such as offsets and cap-and-trade policies, as part of 'the radical Left's progressive wish list'. This is not surprising to many on the left but this also offers much food for thought for students of policy analysis, who focus rather narrowly on coming up with 'efficient solutions'. While smart analyses can be helpful, the belief that such analyses are sufficient to drive policy change has proven to be a chimera. This is why it is necessary to cultivate a strong awareness of the extent to which the beneficiaries of the status quo use their political and economic power to thwart sensible debates on the unprecedented environmental and social crises.

This edited volume is meant as an early intervention to bring consideration of social, economic, and environmental justice to the centre of climate change debates in India. Considering the vastness of the subject matter at hand, there seems to be no better way than to convene a group of fresh voices engaging specifically with each of the many aspects of climate justice. Unlike many other edited volumes, this one is not merely a by-product of a conference or a workshop. The contributors were kind enough to respond to my invitation to write a chapter specifically for this volume. Yet, this was not easy, as this collaboration entailed working through more than one draft of the chapter abstract followed by several drafts of each chapter. Such close and enriching collaboration with the contributors helped produce chapter texts that offer fresh insights at the cutting edge of these pressing debates.

Climate justice debates in India in the past have foregrounded struggles against the strangulating hold of the forces of global capitalism, neocolonialism, and neo-imperialism. These are valid concerns – a frontal response to these regressive forces is necessary to realize a better future for the planet and the majority of the population of the world. Yet, the task of addressing the serious threat that the climate crisis poses to the lives and livelihoods of poor and marginalized groups, including the urban poor, cannot wait for victories against those formidable adversaries. Equally important, the beneficiaries of the status quo continue to seek to mould the global climate policy process and national policymaking processes to suit their interests.

This is why it is risky to focus narrowly on climate advocacy driven solely by the goal of reducing average atmospheric temperatures, no matter how radical the target. Such advocacy is premised on two assumptions that are rarely made explicit: in many instances, aggressive climate action is equated with climate justice. If climate crisis affects the poor and the marginalized the most, wouldn't 'fixing' the climate crisis automatically minimize vulnerabilities and produce climate justice? Or so the argument goes. Unfortunately, any such expectations must be tempered. As the chapters in this volume show, in the pursuit of climate justice, the means matter as much as the ends. A second and related unstated assumption is that we must prioritize climate action before we can pursue climate justice. In the words of Jonathan Logan, one of the founders of Extinction Rebellion (XR) America, 'If we don't solve climate change, Black lives don't matter. If we don't solve climate change now, LGBTQ [people] don't matter ... I can't say it hard enough. We don't have time to argue about social justice.'

Make no mistake, the arguments that the likes of Logan are making are based on an ideology of authoritarian environmentalism. It seeks to use the climate crisis as a totalizing cause to marginalize considerations of a just world. These developments should alert advocates of environmental and climate justice in India. We already have the ingredients necessary for an authoritarian and Malthusian movement on climate action. It is not a coincidence at all that in his novel, *The Ministry for the Future,* Kim Stanley Robinson chose to use India as the site for a hypothetical unilateral deployment of planetary-scale solar geoengineering operations. Deeply entrenched socioeconomic inequalities and a disturbingly widespread acceptance of political authoritarianism are essential ingredients for the rise of climate authoritarianism. Yet, this is not merely a war of wits. Any visions of an alternative world must also outline concrete pathways to translate those visions into reality. In this spirit, this volume seeks to mainstream climate justice within nascent discussions on climate policy and programme development in the Indian context.

Each chapter engages with specific here-and-now issues that sit at the intersection of the climate crisis and socioeconomic crises, of which we have plenty. However, none of the contributors relies on simplistic technocratic solutions that are often presented as silver bullets. Each chapter points to more difficult but enduring tasks of building social, economic, and environmental resilience in sectors as diverse as food, water, energy, including coal and the transition to renewable energy, urbanization, and climate policy development at both the national and state levels. None of the contributors expects to see any major changes to occur without powerful grassroots mobilizations coupled with supportive political and policy advocacy. The history of environmental social movements in India offers deep lessons about building more inclusive climate social movements. While each chapter offers a deep-dive into a specific topic, a comparative reading of the chapters offers cross-cutting insights that will help build bridges across sectors.

In curating this volume, I have drawn inspiration from many conversations that helped animate some of the key arguments that appear in this volume. This included a fortuitously timed invitation in April 2020 to address a webinar as part of the aptly named *Solidarity Series: Conversations During Lockdown and Beyond* organized by the Centre for Financial Accountability, New Delhi. A second virtual talk delivered in February 2021 as part of a series on *Anti-Caste Politics and Environmental Justice* co-organized by Seshadripuram Evening Degree College, Bangalore, and Anti-racist Research and Policy Center (ARPC), American University, created a productive space for some deeper thinking on questions of caste-based oppressions and its implications for climate justice. Two grants from the University of Connecticut were vitally important to this process. A Research Excellence Program grant from the Office of Vice President for Research supported travels to India in the summer and winter of 2019. A Human Rights Faculty Seed Grant for my research on *Economic and Social Rights in a Climate-Changed World* allowed me the space in Spring 2021 to conduct the last round of work on the editing and writing for this volume.

The reflections, arguments, and insights that appear in the volume are a result of engagements with many activists, researchers, and scholars. Rahul Banerjee and Soumya Dutta generously shared their rich understanding of the histories of different strands of environmental and climate movements in India. Several conversations with Nagraj Adve, Rajeswari Raina, and others involved in the Teachers Against Climate Crisis group motivated me to expand the circle of engagement for this volume. Navroz Dubash at the Center for Policy Research (CPR) and colleagues at India Climate Collaborative offered valuable support for the widest possible dissemination of the book.

I am grateful to each of the contributors who worked patiently and diligently on this long and sometimes arduous journey and to the artists, poets, and translators for their creative contributions. The photograph on the book cover showing fishing boats parked in a clock-shape in the village of Champu Khangpok, Langolsabi, Loktak Lake, is by photographer Deepak Shijagurumayum. The photograph also represents the spirit of Naamee Lup, which is a banner representing the collaborative work of the NGO Indigenous Perspectives, Fisherfolk Unions, ESG Bangalore, and other civil society organizations. We are thankful to Ram Wangkheirakpam, the Convenor of Indigenous Perspectives, Manipur, for his kind support. Special word of thanks for Eric Chu who provided valuable inputs on numerous occasions. This volume has benefited from the diligent copy-editing work of Chitralekha Manohar and her team at The Clean Copy and a very supportive steering of the editorial process by Anwesha Rana and Qudsiya Ahmed at Cambridge University Press.

The love, care, and discipline that keeps me on track comes from my family in Storrs – Saroj, Zia, and Sophie – who have lent strength and emotional energy every step of the way. I dedicate this volume to the social justice advocates and climate activists who have demonstrated the courage and conviction needed for persevering in the hope of a more just world.

Storrs, CT   **Prakash Kashwan**
April 5, 2022

Empowerment by Nidhin Donald

Chapter 1

Introduction
Climate Justice in India

Prakash Kashwan

Arundhati Roy famously described the COVID-19 pandemic as a

> portal, a gateway between one world and the next. We can choose to walk
> through it, dragging the carcasses of our prejudice and hatred, our avarice, our
> data banks and dead ideas, our dead rivers and smoky skies behind us. Or we can
> walk through lightly, with little luggage, ready to imagine another world. And
> ready to fight for it. (Roy 2020)

As inspiring and insightful as these words are, such juxtaposition of utopia and
dystopia barely scratches the surface of what and who we *are* as a nation. The soul-
crushing images of burning pyres in parking lots turned into makeshift graveyards,
which international and national media have immortalized, offer a clue, as does the
sombre poetry of Parul Khakhar (Tripathi 2021). India is a land pockmarked with
a million fires.

The COVID-19 crisis has come as a shock to many middle-class Indians. Yet, to
India's Dalits, Adivasis, women, and other marginalized groups, haunted by centuries
of oppression, this crisis is yet another in a long list of historical and ongoing crises.
For example, the coalfields of Jharia in Jharkhand have been burning for over a
century now. As a result, at least 130,000 families have, quite literally, lived through
a century-long trial by fire (Rahi 2019). Since 1995, the state-owned Bharat Coking
Coal Limited (BCCL) has claimed to have a 'master plan', which is possibly gathering
dust in some almirah of the coal ministry (S. Kumar 2021). One would imagine that
a pandemic like COVID-19 might scare the minister whose job includes ensuring

the welfare of the 3.6 million people who work in mines with a less than adequate supply of fresh air. Yet, in 2020, India's coal minister valorized coal workers as 'our coal warriors who are toiling day and night to keep the lights on even during the corona pandemic' (Press Information Bureau 2020). They toiled very hard indeed.

A year later, as India struggled to confront the monstrous second wave of the pandemic, Central Coalfields Limited (CCL), a subsidiary of Coal India Limited (CIL), recorded the highest-ever single-day coal dispatch of 80 railway rakes (PNS 2021). Unfortunately, such exceptional productivity in the middle of a pandemic came at a steep cost, as at least 400 CIL employees died from COVID-19. CIL appealed publicly to Prime Minister Narendra Modi, requesting about 1 million doses of vaccines for its employees (Singh 2021). However, it is unclear if CIL's request was fulfilled. Nevertheless, India's coal workers and the residents of Jharkhand, the latter hardened by century-long neglect and violence of extractivism, continue to be caught in the crossfire between advocates of national development and stakeholders in the ongoing contestations over the impending renewable energy transition. The involvement of these varied parties and interests has not translated into negotiating power for mine workers, as seen among their counterparts in the West, who have managed to mobilize under the banner of a just transition.

In the midst of the COVID-19 pandemic, the Government of India 'unleashed coal', that is, they opened up coal mining to the private sector. In doing so, Prime Minister Modi declared that he was 'unshackling [coal mining] from decades of lockdown', as he wanted 'India ... to be a net exporter of coal' (Varadhan 2020). This celebration of coal is linked to long-standing traditions of coal nationalism (Lahiri-Dutt 2016). For the Indian prime minister, the advocacy and support for expanding coal mining does not appear to conflict with the country's ambition of playing a prominent role in global climate negotiations. At the Leaders Summit on Climate convened by United States President Joe Biden, Modi announced the US–India Clean Energy Agenda 2030 Partnership, which is to 'proceed along two main tracks: the Strategic Clean Energy Partnership and the Climate Action and Finance Mobilization Dialogue' (CNBC TV18 2021). How might these partnerships and India's continued expansion of coal mining shape India's climate action, and the welfare of the multitude of coal miners, most of whom work under extremely exploitative conditions? What will happen to the young boys descending steep chutes – little more than 'rat holes' – to dig coal from hard rock, with just a pickaxe and a torch, in the Jaintia Hills in eastern India (Chandran 2016)?

These snapshots from the year of the pandemic help to outline how Indian leaders respond to crisis situations. They also offer a glimpse of what a major and widespread crisis portends for the majority of India's people, whose lives are locked in multiple intersecting circles of crises and immiseration. A consideration of how myriad social, economic, and ecological crises reinforce the vulnerabilities experienced by the most marginalized, and their efforts to overcome those vulnerabilities, should be at the heart of the pursuits of climate justice.

Climate change in a grossly unequal society

The climate crisis is occurring in a world of extreme inequalities. The history of disproportionate contributions to the accumulation of greenhouse gases (GHGs) responsible for the current crisis is truly staggering. As of 2019, a handful of countries, including the United States, United Kingdom, countries of the former Soviet Union, Germany, France, Poland, Canada, and Japan, contributed about 75 per cent of the world's historically accumulated emissions. China alone was responsible for about 18 per cent. The majority of the world's countries collectively contributed only 7 per cent to the total GHG emissions present in the atmosphere today. These inequalities would be even more significant if one were to account for the transfer of consumption emissions via international trade or travel. India has contributed less than 3 per cent to the accumulated emissions (Ritchie 2019). Despite contributing a negligible share to the accumulated stock of GHGs, various global indices rank India among the countries most vulnerable to the effects of the ongoing climate crisis (Reuters 2018). As such, India is a victim of international injustices associated with the climate crisis.

India is also home to the largest population of poor people anywhere in the world and is one of the most unequal countries globally today. Ranked according to the Gini coefficient, a national-level measure of inequality in income distribution, India was second only to Russia as of 2018 (Chaudhuri and Ghosh 2021). Concepts such as income inequality and poverty do not quite capture the deep-seated nature and wide-ranging effects of caste-based oppressions. Dalit men are lynched for falling in love with non-Dalit women, and Dalit women are routinely raped with impunity. India's National Crime Records Bureau (NCRB) reported that 10 Dalit women were raped daily in 2019 (Kumar 2020). Even more worryingly, Dalit women are often 'raped to keep them "in their place"' (Nagaraj 2020). The disadvantages that Dalit women face are a product of the oppressive caste system and patriarchal norms at home and in the society at large. The oppression of Dalit men and women is instrumental to the power, authority, and privileges upper-caste men enjoy in India. Caste hierarchy is therefore an embodiment of violent social norms with widespread social acceptance in today's India (Coffey et al. 2018).

Considering these challenges, the editor and contributors to this volume have grappled with how best to refer to a normatively repelling social reality in which many Indians consider references to 'lower' caste and 'upper' caste as objective descriptions. Caste is socially constructed and therefore always political, even when discussed in other contexts. In this text, we will use the vocabulary of 'upper caste' and 'lower caste' to designate groups of people, their experiences, and how they are represented in public discourse. The quotation marks here indicate our personal disavowal of this system of caste hierarchy and its continued normalization in public discourses and writings.[1] But for the sake of brevity, we use these phrases without scare quotes in the remainder of this volume.

The nexus of the climate crisis and socioeconomic and political inequalities is at the root of various types of climate injustices. For decades, hundreds of thousands of poor Indians have died prematurely because of unacceptably high levels of air and water pollution. A recent study estimates that about 2.5 million people in India die every year because of toxic air (30.7 per cent of all deaths in the country) (Vohra et al. 2021). Similarly, the tens of millions of people displaced by annual floods, the hundreds of deaths because of heatwaves, and enormous disruptions to poor people's lives due to climate disasters find scant mention in the national press. These statistics are rarely a subject of public debate in India, except when a health minister, who also happened to be a doctor, denied the existence of data that link air pollution to premature deaths in India (Kaur 2019). Clearly, the worst impacts of air pollution and the climate crisis are being denied, ignored, and normalized, because these burdens fall on the urban poor, women, Dalits, Adivasis, Muslims, and other marginalized people with little political voice. Accordingly, India is an archetypal site for the manifestation of the myriad injustices associated with the climate crisis.

The COVID-19 pandemic has further exacerbated India's inequality problem. The catastrophic failure to plan for the widely anticipated second wave of the COVID-19 pandemic exposed the dark underbelly of India's public institutions, and the lack of freedoms afforded to the press and civil society (Ghoshal and Das 2021). In 2020 alone, an additional 75 million people in India were pushed into poverty, accounting for nearly 60 per cent of the global increase in poverty that year (Lee 2021). In the same period, India counted 55 new billionaires, or about one billionaire every week, despite a major economic slowdown in the wake of the hastily declared and rashly managed nationwide lockdown (Bhargava 2021).

Unequal societies are badly governed – they do not have what it takes to rein in the exploitative and polluting models of extractive development that corporations and political-economic elite find beneficial and perpetuate. A careful reading of the

---

[1]   I am grateful to Srilata Sircar for this formulation.

available scientific evidence would suggest that inequality, not poverty, is the biggest polluter (Oxfam International 2020).

Failure to remedy environmental degradation and stabilize the global climate system aggravates these injustices; yet not all environmental and climate action addresses injustices. Paradoxically, many types of interventions meant to mitigate the impacts of climate change are likely to further reinforce these pre-existing inequalities. As this volume goes to press, 1,500 families in central Assam's Nagaon district are fighting to regain control of 276 *bighas* (a varying measure of land area used in India and other parts of south Asia) of farmland forcibly acquired for a 15-MW (megawatt) solar plant being developed by Azure Power Forty Private Limited. According to a group of over 150 academics, activists, lawyers, students, filmmakers, and other concerned citizens, the land acquisition process in this case violates Assam's land laws as well as the residents' human rights (*The Hindu* 2021). Similar injustices are likely to repeat all over the country, as India plans to rely on the expansion of solar and wind power to achieve its intended nationally determined contributions (INDCs) to the Paris Climate Agreement. However, if not handled with the utmost care, this keenly anticipated renewable energy revolution could add significantly to India's long-standing and worsening land wars (Levien 2013).

To those focusing on radical climate action, the injustices resulting from such action may seem mere aberrations. Indeed, in the Global North, where debates surrounding climate justice have been around for longer, some scholars and activists equate radical climate action to climate justice (cf. Kashwan 2021). However, the climate crisis, climate denialism, and the dismal outcomes of international climate negotiations share the same roots: the influence of exploitative and extractive systems of global capitalism, which are propelled by a nexus of multilateral financial institutions and national political and economic elites. The power of this loosely organized, yet extremely nimble, web of transnational elite networks is rooted in histories of colonialism, imperialism, and neocolonialism (Bachrach and Baratz 1962). Activists and scholars focusing on global capitalism have paid inadequate attention to how such networks thrive on intersectional inequalities borne of the confluence of gender, caste, class, and religious identities within countries. To this day, these inequalities help forge social relations, institutional arrangements, and political structures that shape socioeconomic, environmental, and policy outcomes. Furthermore, the climate crisis greatly exacerbates these inequalities and injustices.

Climate Justice in India is the first comprehensive book-length effort to examine how the climate crisis and some of the proposed solutions are inextricably linked to social and economic justice in Indian society. In this volume, we push back against climate policy discussions that deprioritize questions of inequalities and injustice, as if they can be addressed post facto. Some policymakers and policy experts assume that the agenda of climate justice has potentially negative consequences for India's international negotiating positions (Swarnakar 2019). However, such nationalism rings hollow. It is evident that no nation can thrive, internationally or locally, without ensuring the well-being of all of its people, environment, and ecology.

Analysing the policies and politics of climate action is the necessary first step to preventing vested interests from derailing meaningful progress in climate action and climate justice. Yet better data or improved analyses of how to 'balance' the considerations of climate action with those of climate justice are unlikely to be sufficient to bring about such a change. Decades of social science evidence suggests that meaningful institutional, political, and economic reforms that serve the interests of marginalized groups like Adivasis, Dalits, and women cannot be accomplished without formidable social and political mobilization (Kashwan 2017). With this in mind, we articulate a politically conscious approach to climate justice that draws on social scientific theories suited to an analysis of the socioeconomic and political realities of India. We take the histories of colonialism and the realities of neo-imperial capitalist capture seriously; we also avoid post-modernist abstractions that fail to address the role of specific actors and agencies in producing climate vulnerabilities at the global, national, and sub-national levels. Moreover, since the beneficiaries of the status quo pursue their agendas by taking over political and policy processes, we need a forceful engagement with these processes to reclaim power from extant regimes.

Through the chapters in this volume, we make five key contributions to the ongoing debates and nascent scholarship on climate justice in India. One, we advance debates on climate justice beyond the long-standing stalemate between questions of international climate justice and the grave domestic inequalities that climate change is likely to greatly exacerbate. For instance, we examine the contents of national- and state-level climate action plans, analyse the evolution of urban climate governance and investigate the relationship between economic inequality and state-level carbon emissions. Two, we bridge the ever-present gap between critical social science scholarship and largely technocratic, apolitical policy-oriented writings. We employ historically informed, empirically grounded, and conceptually rich social science analyses to inform policy and programmatic debates about climate justice in India. For example, in two chapters, we apply the concept of intersectionality to investigate how gender- and caste-based inequalities together influence access to drinking water and the outcomes of agroecological farming.

Three, we seek a carefully curated balance between conceptual richness and the sectoral and contextual specificity of the varied manifestations of climate injustice in both rural and urban India. This includes discussions on inequalities in carbon emissions, energy justice, natural resource extraction, gender- and caste-based determinants of access to clean drinking water and agroecological farming, urban climate justice, climate movements, and analyses of national and state climate action plans using a climate justice lens. Four, our contributions are grounded in a deep understanding of the Indian context, but each chapter also speaks more broadly to themes prominent in debates on climate justice in other countries of the Global South. Five, the contributions to *Climate Justice in India* reflect a philosophy of theoretical, methodological, and epistemological pluralism.

In the next section, I offer information essential to understanding the historical and more recent causes of the climate crisis. The third section contains a broad framework for climate justice, which formed the basis of my editorial engagement with the volume's contributors. In this framework, I complement the key constituent elements of justice, as argued by justice theorists, with a focus on political and policy processes needed to bring about transformative change. Analyses of policies and policy processes include thinking through the workings of intersectional inequalities given India's social, economic, and political contexts. In the final section, I offer a broad overview of the major ongoing debates on climate justice and, accordingly, situate individual contributions to this volume.

Background: Colonial and post-colonial sources of climate vulnerability

The most common conceptualizations of climate justice speak of an uneven distribution of the costs and burdens of the ongoing climate crisis along axes of nationality, ethnicity, gender, sexuality, caste, and class, among others. These are the distributional aspects of climate justice. Other important dimensions of justice include procedural, recognitional, and reparational work. A systematic analysis of the historical, political, and economic contexts of the genesis and development of the ongoing climate crisis is indispensable to a nuanced understanding of the contemporary manifestations of injustice and the pursuit of climate justice.

Colonization, imperialism, and capitalism

Colonialism is the domination and subjugation of a people by another, most commonly the settler and non-settler European colonization of the Americas, Australia, and parts of Africa and Asia (Kohn and Reddy 2017). Colonial rule led to massive extractions of natural resources and the rampant exploitation of people in the colonies to serve imperial expansion. The mobilization of the unpaid labour of colonized and enslaved people for the production of 'cheap nature' were central to 'the endless accumulation of capital' (Moore 2016, 79). Economist Utsa Patnaik estimates that between 1765 and 1938, the East India Company and the British Raj siphoned off at least £9.2 trillion ($44.6 trillion) worth of unaccounted wealth (Sreevatsan 2018). Patnaik also shows that the combined drain from Asia and the West Indies constituted about 6 per cent of Britain's gross domestic product (GDP) from 1780 to 1820, a crucial period in its industrial transition.

The processes of colonialism and capitalism shaped the political-economic system that emerged in the postcolonial era. This included the Bretton Woods Institutions, that is, the World Bank and the International Monetary Fund founded in 1944. Gross inequalities in international economic, trade, and financial systems enable the continued exploitation of resources and people on the periphery and fuel patterns of wasteful and profligate consumption in the Global North. These patterns of resource use drive the exploitation of the global atmospheric commons, which act as sinks for GHGs from industrially advanced countries (Bassey 2012). However, the legacies of colonization extend far beyond material exploitation. Colonialism deepened the feudal tendencies inherent in Indian society and weaved caste hierarchies into political and institutional structures. Such institutionalization of social and political hierarchies initiated processes of internal colonialism, in which large sections of populations within formerly colonized states were colonized by their own ruling elite, often acting in the name of 'development' (Calvert 2001, 51). More broadly, the present-day social, cultural, psychological, political, economic, and institutional effects of colonialism are equally important (O'Dowd and Heckenberg 2020).

Let me cite three examples to illustrate the contemporary effects of colonialism and the postcolonial politics of resource control. One, policies related to the management of natural resources that rely on forest–farm distinctions draw on caste–tribe differentiations that were present in precolonial India but solidified significantly under colonial rule. These distinctions supported resource extraction regimes that were crucial to the colonial project and continue to shape contemporary models of forest governance, regimes of forest rights, and the extraction of valuable minerals, which fuels domestic and global capitalism (Kashwan 2017). Two, the development of the ecologically fragile northeast India as the country's hydropower hub is a direct result of New Delhi's political dominance, long-standing patterns of uneven regional development, and a reliance on top-down models of development and governance in 'a racialized frontier region' (Gergan 2020, 1–2). Three, most Indian cities were designed with the dual goals of facilitating assorted trade and commerce and protecting the health and wealth of a small population of colonial elite, while pushing the majority of urban populations to the margins. For example, colonial town planners, financiers, and property developers collectively secured Bombay as a space for commerce by categorizing different types of neighbourhoods as legitimate or illegitimate (Chhabria 2019). This helped 'delimit the city as a distinct object and progressively exclude laborers and migrants, who were forced into the so-called "slums"' (Chhabria 2019). The colonial-era patterns of class-driven differentiation are also evident in present-day Mumbai (Farooqui 1996; Bhide 2015).

These examples are meant to illustrate specific outcomes that are rooted in and reinforce well-entrenched social, economic, and political inequalities. The patterns of pervasive disparities common to settler colonial societies of the Americas are also present in India, such as in the discriminatory and subjugated incorporation of the states and peoples of northeastern India (Noni and Sanatomba 2015). Additionally, internal colonization also manifests via caste- and tribe-based inequalities in every sphere of the economy, society, and politics (Desai and Dubey 2011). Routine and generalized policies and programmes cannot address such deep-seated inequalities, which requires deeper engagement.

Caste-, tribe-, and ethnicity-based discrimination

Adivasi communities are distributed across regions rich in forests and other natural resources; this has made them targets of land grabs, resource grabs, and green grabs, that is, taking control of a territory in the name of environmental conservation (Kashwan, Kukreti, and Ranjan 2021). Similarly, Dalits and Muslims have been subjected to political and economic control by beneficiaries of the status quo, primarily people from the higher castes (Dey 2019). The pervasive nature of such inequalities is evident in the fact that Dalits, Adivasis, and Muslims are under-represented at the highest levels in nearly every sector of society, including the press, cinema, science, higher education, and political leadership. Some scholars argue that the emphasis in social science research on 'the binary of colonialism versus nationalism' is why Dalits and their questions have been missing from academic knowledge production in India (Rawat and Satyanarayana 2016, 9). The existence of internal colonialism and these deeply entrenched inequalities has grave implications for environmental and climate vulnerabilities.

Take, for example, the widely discussed topic of air pollution. It is well known that exposure to air pollution depends on class position – the poor are exposed to the worst forms of pollution for the longest duration in a 24-hour cycle (Wu et al. 2020). Yet 'class' is only one of the many dimensions of inequality and discrimination that is relevant to the production of vulnerabilities. Gender is another important determinant of disadvantage. A study by the Council on Energy, Environment and Water (CEEW) shows that household heating and cooking accounted for 40 per cent of the pollution in Delhi in December 2020 and January 2021 (*Livemint* 2021). Indeed, the burden of household chores falls disproportionately on women, who experience the most direct impacts of indoor air pollution in both urban and rural settings.

The enormity of the problem becomes apparent when one accounts for the cross-cutting effects of caste, class, gender, and religion. Addressing such intersectional disadvantages requires broad interventions and transformative change in the social, cultural, economic, and political spheres. Climate crisis exacerbates the effects of pre-existing inequalities. Moreover, the pervasive nature of multiple inequalities blunts public demand for more egalitarian policies (Melo, Ng'ethe, and Manor 2012). Clearly, the pursuit of climate justice is a daunting challenge. However, attempts to narrow the definition of climate justice are unhelpful. Climate justice simply cannot be separated from broader and entrenched socioeconomic and political inequalities.

Climate justice: a conceptual framework

The vastness and complexity of the climate justice agenda necessitate the use and development of theories and insights from multiple disciplines. Of course, interdisciplinary and collaborative discussions and interventions among researchers, activists, and policymakers require all participants to be familiar with the basic tenets of justice theory and how these may be combined with insights from the social and natural sciences. In the absence of such engagement, as Lianghao Dai argues, we risk promoting fake interdisciplinary collaborations (Dai 2020).

In this section, for a more comprehensive understanding of climate justice and its manifestations, I introduce concepts foundational to justice theory. These include the three constituent elements of justice – distribution, procedural, and recognition – which justice theorists use frequently. Towards the end of this section, I discuss two additional aspects – restitution and reparation – that have entered climate justice debates relatively recently.

Distributional justice refers to the fair distribution of the costs and burdens of climate change and societal responses to it. As mentioned previously, climate change responses create opportunities for some, and costs and burdens for others. Carbon offset projects, in which industrial giants and multinational corporations 'compensate' for their emissions by funding forest conservation projects in the Global South, have led to the violent dispossession of indigenous and other forest-dependent people (Kashwan 2015; Ghosh 2020). Researchers refer to these and other projects that seek to recompense for industrial emissions and consumerist lifestyles in the Global North as instances of carbon colonialism (Agarwal and Narain 1991). Procedural justice is about whether the groups most affected by climate change have adequate opportunities and the means to engage in the brainstorming, design, and implementation of climate policies and actions. Recent scholarship urges us to look beyond the distributional and procedural dimensions to examine whether marginalized groups are recognized as legitimate claimants and stakeholders in relevant political and policy processes, and if their experiences of the costs and risks of climate change inform the design of policies and programmes meant to advance climate action (Schlosberg and Collins 2014; Chu and Michael 2019).

Each of the three dimensions of climate justice can be applied to one or more of the following areas of climate change policy and research: climate mitigation, climate adaptation, and climate resilience. Climate mitigation includes actions aimed at reducing and eliminating GHG emissions. Climate adaptation refers to the measures intended to minimize the impacts of climate change, some of which may help reduce vulnerabilities to the future effects of climate change.

The failure of the international community and national government to ensure just climate mitigation and adaptation interventions means that ongoing climate change imposes unmitigated burdens and costs on poor and marginalized groups. Many of these impacts have been studied through the lens of climate resilience, which draws attention to anticipatory interventions meant to strengthen communities' abilities to withstand the effects of climate change (Kim, Marcouiller, and Woosnam 2018). However, in some cases, the concept of 'resilience' has been used to focus too narrowly on the actions and strategies of vulnerable communities, without accounting for the structural forces of colonialism, patriarchy, and casteism, which are responsible for communities' lack of resilience or high vulnerability (Cote and Nightingale 2012; Kashwan and Ribot 2021).

The intersection of the two analytical planes discussed here – three constituent elements of justice (distributional, procedural, and recognition) and three aspects of climate change (mitigation, adaptation, and resilience) – yields a useful scaffolding for understanding climate justice. While these dimensions are the mainstay of much past academic work and activism, recent debates recognize the importance of two other dimensions: restitution and reparation.

Restitution refers to the restoration of something – often lost or stolen – to its rightful owner. For example, lands and territories that settler colonial, national governments, or other dominant social groups took away from indigenous and other rural communities, thereby creating a class of dispossessed peasants. Rectifying these past injustices requires the restitution of 'access to land, territory, water, forests, especially in light of the global land grabbing during the past decade' (Borras and Franco 2018, 1319). In the context of the climate crisis, philosophers argue that some actors, for example, fossil fuel corporations and the countries of the Global North, which are responsible for the climate crisis, owe restitution to those most affected by it (Gardiner 2011). This principle informs the demands of countries in the Global South, that industrially advanced countries pay for the loss and damages linked to the climate crisis. Indeed, such demands could also be applied within national borders. In India, this relates most directly to the restitution of land, forest, and other resource rights to Dalits and Adivasis, who suffer high rates of landlessness and criminalization of resource use because of state control of resources.

Demands for the protection of resource rights and restitution of lost lands are codified in acts of Parliament, such as the Panchayat (Extension to Scheduled Areas) Act (PESA), 1996, and the Forest Rights Act, 2006 (FRA). However, the state has failed to implement these laws because they threaten the undue advantage that powerful actors in the state and society enjoy in the status quo. For example, as of May 2021, 40 per cent of states had not formulated the rules necessary for the implementation of PESA (Pandey 2021). Unfortunately, the lingering effects of the caste–tribe dichotomy and instrumental use of the narratives of Adivasi rights towards forest protection have led to a neglect of Dalit land restitution (Prasant and Kapoor 2010). Moreover, Dalits have also been victims of the enclosure of village commons by forest departments throughout the country (see Table 3.1 in Kashwan 2017, 58). Such appropriation and continued occupation of village commons violate the United Nations Declaration on the Rights of Peasants and Other People Working in Rural Areas (UNDROP), adopted by the UN General Assembly in October 2018 (Kashwan, Kukreti, and Ranjan 2021).

The provisions of UNDROP apply to Dalits and other landless rural workers too. Unfortunately, domestic debates about land reform and redistribution to Dalits have never really taken off because of mainstream Hindu society's delegitimization of Dalits as agriculturalists (Rawat 2011). The marginalizing and invisibilizing of Dalit land claims continue in neoliberalized India today; some even argue that land dispossession exacted in service of 'new economy projects may be liberating for Dalits' (cf. Agarwal and Levien 2020, 696). The promise that neoliberal economic reform will bring prosperity to the poor is yet to be fulfilled, in part because these reforms have never really articulated and incorporated the interests of poor people. On the contrary, the corporate control of the economy and free flow of speculative global finance have led to the selective withering of the welfare state and the militarization of the state's appropriation of land and natural resources (Ram 2012; Agarwal and Levien 2020).

All of these outcomes are because of the multiple and concentrated disadvantages that Dalits, Adivasis, the northeast tribes, and Muslims face in a neoliberalized India. These groups lack representation in the public sphere–they are unable to shape public agendas, they are excluded from political and policy processes, and they lead precarious lives because of their high income and wealth poverty. The neoliberal reset of the welfare state, and capture of the political agenda by advocates of global capitalism in India and elsewhere, work through these debilitating inequalities and exclusions (Kashwan, MacLean, and García-López 2019). This is why there is little sustained and informed public debate on the alarming levels of pollution in Indian cities, the dangerously high fluoride content of drinking water in many parts of the country, and extreme disparities in access to safe sanitation (Chaudhuri and Roy 2017). These background conditions make a huge percentage of India's population highly vulnerable to climate shocks and stresses. COVID-19 exposed the glaring forms of exclusion and marginalization that the urban poor, especially migrant workers, face (Suresh, James, and Balraju 2020). Advocates of climate justice need to grapple with these long-standing inequalities present in every nook and cranny of India's vast and complex rural and urban geographies.

Overview of the chapters and their debates

India is a land of competing inequalities; it presents a challenge to researchers of inequality and justice. If the devastating images of COVID-19 are any indication, urban India is likely to be a climate justice hotspot in the near future. The UN estimates that between 2018 and 2050, India will have 416 million new urban dwellers (UN-DESA 2018). Such rapid urbanization will put significant pressure on rural and forested areas, which are the sources of natural resources needed for urban infrastructure development and the sustenance of large urban populations.

The nature of urban growth and manner of urban climate mitigation and adaption planning and execution have significant implications for urban climate justice (Shi et al. 2016). Eric Chu and Kavya Michael take on this challenging topic in Chapter 2; they analyse ongoing interventions related to urban climate adaptation, risk reduction, and resilience-building actions. However, instead of adopting a narrow programmatic focus, they situate these developments within the country's recent history of neoliberal economic transformation and long-standing socioeconomic inequalities. Although Indian leaders identify local development priorities as the main entry point for climate mitigation and adaptation in India's cities, market actors often assume control of these opportunities to the exclusion of the majority of urban populations (Khosla and Bhardwaj 2019).

An equally important area of focus is the much-anticipated transition to renewable energy, which has prompted a vigorous scholarly and policy debate on energy justice in the Global North (Sovacool et al. 2017). Yet there has been little work on this transition in the Indian context (Yenneti and Day 2015). In Chapter 3, K. Rahul and Parth Bhatia fill this gap by exploring the benefits and challenges of adopting energy democracy and energy justice. They look at three types of renewable energy developments in India: large-scale renewable energy projects, solar pump sets, and energy access programmes. In India, however, the framework for a just transition has been criticized from the perspective of the context and vulnerabilities of workers employed in mining and various other operations of the fossil fuel industry (Roy, Kuruvilla, and Bhardwaj 2019). Still, the majority of people employed in the sector work under exploitative and environmentally hazardous conditions that are common to India's coal industry (Lahiri-Dutt 2016).

Recent work has enhanced our understanding of the political economy of India's extractive regime (Adhikari and Chhotray 2020). In Chapter 4, Vasudha Chhotray builds on her field research in Jharkhand and Chhattisgarh to expand the scope of just transition research beyond labour; she situates it within broader political and economic systems with high levels of inequalities. Chhotray also highlights the multifaceted spaces that social and climate justice activists could mobilize for a just transition.

Ensuring justice in the ongoing transition is not easy, especially because of the pervasive changes in the economy and politics. Haimanti Bhattacharya offers one example of a major pervasive change in Chapter 5. Based on her recent and ongoing research, Bhattacharya shows that the relationship between carbon emissions from fossil fuels and inequality in consumption expenditure at the state level has undergone a major transformation since the onset of the economic reforms in 1991. Bhattacharya's findings reinforce the proposal other scholars have made in favour of a carbon tax, based on household consumption, and that such taxes should be utilized to pursue broad-based goals of energy and transportation justice (Azad and Chakraborty 2020). Similar policies in other sectors of the economy should be the focus of India's climate strategy. Unfortunately, such a policy focus is missing from India's national and state climate action plans, as Arpitha Kodiveri and Rishiraj show in Chapter 6. They review India's national and state climate action agendas to determine if and how they incorporate concerns of climate justice.

Despite India being among the most vulnerable countries, the Indian Parliament has not even debated, let alone enacted, a climate change law. Instead, India's climate change responses are governed by various executive orders and ad-hoc climate action plans; this is a cause for concern. Quite tellingly, the country's first climate change bill was a private member's bill that influential Bharatiya Janata Party leader Jayant Sinha introduced in March 2021. This bill seeks to provide a framework 'by which India can develop and implement clear and stable climate change policies' under the Paris Climate Agreement (Farand 2021). This is an intriguing proposal coming from a member of parliament (MP), who represents the coal-producing Hazaribagh district in Jharkhand; this illustrates the complexity of politics over climate strategy. The justice implications of these developments are quite significant. Emissions from the ongoing burning of fossil fuels and profligate consumption by a rapidly growing Indian elite class must then be offset by planting forests, modifying agriculture and other land-use patterns, or resorting to other carbon dioxide removal techniques. Net-zero plans essentially transfer the burdens of climate action between different sectors of the economy, for example, when industrial emissions are sought to be offset by planting trees in village commons (Skelton et al. 2020). In essence, the nascent plans for India's climate response are rife with potential for domestic injustices of numerous types.

None of this is new. As I show in Chapter 7, many of India's climate activists have been warning of these possibilities since the early years of the new millennium. That said, I argue that a fuller appreciation of the complex challenge of social mobilization for climate justice requires a deeper understanding of the history of environmental movements and the debate on the varieties of environmentalisms in India. To this end, I investigate three of the most successful environmental movements in India and highlight the implications of the multi-scalar nature of both environmental and climate movements and their engagements with mainstream political spaces. These analyses shed light on the trajectories of arguments about international and domestic climate justice in India, and the promise of India's nascent climate youth movements. However, it is important to grapple with myriad ways in which social inequalities shape Indian environmental movements (Sharma 2012).

In Chapter 8, Srilata Sirkar poses the unspoken caste question in India's environmental and climate debates. Echoing similar demands about attending to questions of racial justice in the United States and building on recent work conducted in India, Sirkar asserts that caste justice is climate justice. She makes a strong case that India's climate movement needs to be an anti-caste one (Ranganathan 2022). Normative visions of the type Sirkar articulates offer important points of departure for redrawing policies, programmes, and strategies that are necessary for realizing climate justice.

Until this volume, there has been a notable and near-total silence on caste and the impact that climate change may have on Dalits in India (Onta and Resurreccion 2011). However, gender has been the focus of quite a bit of research on climate adaptation recently (Rao et al. 2019). The last two chapters conduct explicitly intersectional analyses of the joint effects of gender- and caste-based inequalities on access to safe drinking water, agriculture, and, more broadly, climate action. In Chapter 9, Vaishnavi Behl and I explore how the intersections of gender-, caste-, and class-based inequalities shape access to clean drinking water in the Garhwal Himalayas and Gujarat. Intersectional injustices also permeate climate adaptation and resilience interventions implemented by multilateral donor agencies and well-known non-governmental organizations (NGOs). We point to the intractable nature of caste and gender inequalities and the limitations of addressing them through programmatic interventions, for example, in the much talked about UN Sustainable Development Goals (Patnaik and Jha 2020). These debates invite climate justice scholars and activists to engage with questions of transformative societal change (Rao and Kelleher 2005; Nightingale et al. 2020).

In Chapter 10, Ashlesha Khadse and Kavita Srinivasan apply the lens of intersectional agrarian justice to analyse ongoing policy and programmatic initiatives meant to promote agroecology, with an emphasis on securing women farmers' land rights (Borras and Franco 2018). These authors apply the framework of intersectional agrarian justice to investigate state-level policies and programmes in Tamil Nadu and Kerala, including the role of women's organizations. Moreover, they use intersectionality to explain what policy and programmatic interventions are likely to work best. In the end, they argue in favour of a hybrid approach that integrates the goal of securing women's land rights with the state effort of promoting agroecology interventions – each is indispensable to advancing intersectional agrarian justice. Their research calls attention to themes of agrarian climate justice and food sovereignty (Agarwal 2018).

In the concluding Chapter 11, Eric Chu and I summarize the key insights from the volume to facilitate broader conversations on climate justice in India and beyond. We reflect on the importance of unifying the diverse voices of academics and social activists engaged in researching various sectoral manifestations of climate governance and climate justice in India. Looking ahead, we outline an engaged research and scholarship agenda that advances academic debate while contributing to the praxis of climate justice. We join others before us in calling for a move beyond the old debates about international versus domestic climate justice to examine the complex intersections of international and sub-national policies, programmes, and resource mobilizations that shape the outcomes of climate action and climate justice (Schlosberg and Collins 2014; Routledge, Cumbers, and Driscoll Derickson 2018; Dubash 2019). Furthermore, we argue for an increased focus on domestic political engagements, accompanied by support and mobilization of transnational human rights and climate justice networks (Kashwan, Kukreti, and Ranjan 2021). Ultimately, though, social mobilizations and political engagements within India are likely to be the major determinants of climate action and climate justice.

References

Adhikari, A., and V. Chhotray. 2020.'The Political Construction of Extractive Regimes in Two Newly Created Indian States: A Comparative Analysis of Jharkhand and Chattisgarh'. *Development and Change* 51 (3): 843–873.

Agarwal, A., and S. Narain. 1991. *Global Warming in an Unequal World: A Case of Environmental Colonialism*. New Delhi: Centre for Science and Environment.

Agarwal, B. 2018.'Can Group Farms Outperform Individual Family Farms? Empirical Insights from India'. *World Development* 108: 57–73.

Agarwal, S., and M. Levien. 2019. 'Dalits and Dispossession: A Comparison'. *Journal of Contemporary Asia* 50 (5): 696–722.

Ahmed, S., and E. Fajber. 2009. 'Engendering Adaptation to Climate Variability in Gujarat, India'. *Gender and Development* 17 (1): 33–50.

Azad, R., and S. Chakraborty. 2020. 'Green Growth and the Right to Energy in India'. *Energy Policy* 141 (June). https://doi.org/10.1016/j.enpol.2020.111456.

Bachrach, P., and M. S. Baratz. 1962. 'Two Faces of Power'. *American Political Science Review* 56(4): 947–952.

Bassey, N. 2012. *To Cook a Continent: Destructive Extraction and the Climate Crisis in Africa*. Oxford, UK: Fahamu/Pambazuka.

Bhargava, Y. 2021. 'In 2020, World Added 3 Billionaires Every Two Days, India Added One Every Week.' *The Hindu*, 2 March. https://www.thehindu.com/news/national/india-adds-40-billionaires-in-pandemic-year-adani-ambani-see-rise-in-wealth-report/article33970268.ece (accessed 24 December 2021).

Bhide, A. 2015. *City Produced: Urban Development, Violence, and Spatial Justice in Mumbai*. Mumbai: Centre for Urban Policy and Governance, TISS. http://hdl.handle.net/10625/56490.

Borras, S. M., Jr, and J. C. Franco. 2018. 'The Challenge of Locating Land-based Climate Change Mitigation and Adaptation Politics within a Social Justice Perspective: Towards an Idea of Agrarian Climate Justice'. *Third World Quarterly* 39(7): 1308–1325.

Calvert, P. 2001. 'Internal Colonisation, Development and Environment'. *Third World Quarterly* 22: 51–63.

Chakravarty, S., and M. V. Ramana. 2011. 'The Hiding behind the Poor Debate: A Synthetic Overview'. *In Handbook of Climate Change and India: Development, Politics and Governance*, edited by N. K. Dubash. New Delhi: Oxford University Press, 218–229.

Chaudhuri, D., and P. Ghosh. 2021. 'Why Inequality Is India's Worst Enemy'. *Down to Earth*, 5 March. https://www.downtoearth.org.in/blog/economy/why-inequality-is-india-s-worst-enemy-75778 (accessed 24 December 2021).

Chaudhuri, S., and M. Roy. 2017. 'Rural–Urban Spatial Inequality in Water and Sanitation Facilities in India: A Cross-sectional Study from Household to National Level'. *Applied Geography* 85: 27–38.

Chhabria, S. 2019. *Making the Modern Slum: The Power of Capital in Colonial Bombay*. Seattle: University of Washington Press.

Chu, E., and K. Michael. 2019. 'Recognition in Urban Climate Justice: Marginality and Exclusion of Migrants in Indian Cities'. Environment and Urbanization 31 (1): 139–156.

CNBC TV18. 2021. 'Climate Summit 2021: From PM Modi to US President Biden, Here's What World Leaders Pledged on This World Earth Day'. 23 April. https://www.cnbctv18.com/world/climate-summit-2021-from-pm-modi-to-us-president-biden-heres-what-world-leaders-pledged-on-this-world-earth-day-9034821.htm (accessed 24 December 2021).

Coffey, D., P. Hathi, N. Khurana, and A. Thorat. 2018. 'Explicit Prejudice: Evidence from a New Survey'. *Economic and Political Weekly* 53(1): 46–54.

Cote, M., and A. J. Nightingale. 2012. 'Resilience Thinking Meets Social Theory Situating Social Change in Socio-Ecological Systems (Ses) Research'. *Progress in Human Geography* 36 (4): 475–489.

Dai, L. 2020. 'What Are Fake Interdisciplinary Collaborations and Why Do They Occur?' *Nature Index*. https://www.natureindex.com/news-blog/what-are-fake-interdisciplinary-collaborations-and-why-do-they-occur (accessed 5 March 2022).

Desai, S., and A. Dubey. 2011. 'Caste in 21st Century India: Competing Narratives'. *Economic and Political Weekly* 46: 40–49.

Dey, D. 2019. 'India: The Context of Its Current Internal Colonialism'. *In Shifting Forms of Continental Colonialism*, edited by D. Schorkowitz, J. R. Chávez, and I. W. Schröder. Springer, 249–272.

Dubash, N. K. 2019. *India in a Warming World: Integrating Climate Change and Development*. New Delhi: Oxford University Press.

Farand, C. 2021. 'Indian Lawmaker Submits Private Bill to Achieve Net Zero Emissions By 2050'. *Climate Home News*. 18 March. https://www.climatechangenews.com/2021/03/18/indian-lawmaker-submits-private-bill-achieve-net-zero-emissions-2050/ (accessed 24 December 2021).

Farooqui, A. 1996. 'Urban Development in a Colonial Situation: Early Nineteenth Century Bombay'. *Economic and Political Weekly* 31: 2746–2759.

Gardiner, S. M. 2011. 'Climate Justice'. In *The Oxford Handbook of Climate Change and Society*, edited by J. S. Dryzek, R. B. Norgaard, and D. Schlosberg. New York, Oxford University Press, 309–322.

Gergan, M. D. 2020. 'Disastrous Hydropower, Uneven Regional Development, and Decolonization in India's Eastern Himalayan Borderlands'. *Political Geography* 80 (June). https://doi.org/10.1016/j.polgeo.2020.102175.

Ghosh, S. 2020. 'Moving Away from State and Capital: Climate Change, Hegemony and Resistance in Indian Forests'. In *Climate Justice and Community Renewal*, edited by Brian Tokar and Tamra Gilbertson. Abingdon, Oxon; New York, NY: Routledge, 51–69.

Ghoshal, D., and K. Das. 2021. 'EXCLUSIVE: Scientists Say India Government Ignored Warnings Amid Coronavirus Surge'. Reuters, 1 May. https://www.reuters.com/world/

asia-pacific/exclusive-scientists-say-india-government-ignored-warnings-amid-coronavirus-2021-05-01/ (accessed 24 December 2021).

Kashwan, P. 2015. 'Forest Policy, Institutions, and REDD+ in India, Tanzania, and Mexico'. *Global Environmental Politics* 15(3): 95–117.

———. 2017. *Democracy in the Woods: Environmental Conservation and Social Justice in India, Tanzania, and Mexico, Studies in Comparative Energy and Environmental Politics*. New York: Oxford University Press.

Kashwan, P., and J. Ribot. 2021. 'Violent Silence: The Erasure of History and Justice in Global Climate Policy'. *Current History* 120: 326–331.

Kashwan, P., I. Kukreti, and R. Ranjan. 2021. 'The UN Declaration on the Rights of Peasants, National Policies, and Forestland Rights of India's Adivasis'. *International Journal of Human Rights* 25 (7): 1184–1209.

Kashwan, P., L. M. MacLean, and G. A. García-López. 2019. 'Rethinking Power and Institutions in the Shadows of Neoliberalism (An Introduction to a Special Issue of *World Development*)'. *World Development* 120: 133–146.

Kaur, Banjot. 2019. 'Minister Claims No Data for Pollution-Related Deaths in India, Studies Say Otherwise'. *Down To Earth*, 4 January. https://www.downtoearth.org.in/news/environment/minister-claims-no-data-for-pollution-related-deaths-in-india-studies-say-otherwise-62706 (accessed 24 December 2021).

Kim, H., D. W. Marcouiller, and K. M. Woosnam. 2018. 'Rescaling Social Dynamics in Climate Change: The Implications of Cumulative Exposure, Climate Justice, and Community Resilience'. *Geoforum* 96: 129–140.

Khosla, R., and A. Bhardwaj. 2019. 'Urbanization in the Time of Climate Change: Examining the Response of Indian Cities'. *WIREs: Climate Change* 10(1). https://doi.org/10.1002/wcc.560.

Kohn, M., and K. Reddy. 2006. 'Colonialism'. *The Stanford Encyclopedia of Philosophy*. Fall 2017 edition. https://plato.stanford.edu/entries/colonialism/ (accessed 24 December 2021).

Kumar, R. 2020. 'On an Average, India Reported 10 Cases of Rape of Dalit Women Daily in 2019, NCRB Data Shows'. News18, 3 October. https://www.news18.com/news/india/on-an-average-india-reported-10-cases-of-rape-of-dalit-women-daily-in-2019-ncrb-data-shows-2930179.html (accessed 24 December 2021).

Kumar, S. 2021. 'Threatened by Fire and Displacement, People of Jharia Dig in Their Heels'. *India Today*, 30 June. https://www.indiatoday.in/india/story/threatened-by-and-displacement-people-of-jharia-dig-in-their-heels-1821090-2021-06-30 (accessed 24 December 2021).

Lahiri-Dutt, K. 2016. *The Coal Nation: Histories, Ecologies and Politics of Coal in India*: New York: Routledge.

Lee, Y. N. 2021. 'Covid Pandemic Pushes 75 Million More People into Poverty in India, Study Shows.' CNBC, 19 March. https://www.cnbc.com/2021/03/19/covid-pandemic-pushes-75-million-more-people-into-poverty-in-india-study.html (accessed 24 December 2021).

Levien, M. 2013. 'The Politics of Dispossession Theorizing India's "Land Wars"'. *Politics and Society* 41(3): 351–394.

Melo, M. A, N. Ng'ethe, and J. Manor. 2012. *Against the Odds: Politicians, Institutions and the Struggle against Poverty*. New York: Columbia University Press.

Moore, J. W., ed. 2016. *Anthropocene or Capitalocene? Nature, History, and the Crisis of Capitalism.* Oakland, CA: PM Press.

Nagaraj, A. 2020. 'India's Low-caste Women Raped to Keep Them "In Their Place"'. Reuters. 25 November. https://www.reuters.com/article/us-india-rape-caste/indias-low-caste-women-raped-to-keep-them-in-their-place-idUSKBN28509J (accessed 24 December 2021).

Nightingale, A. J., S. Eriksen, M. Taylor, T. Forsyth, M. Pelling, A. Newsham, et al. 2020. 'Beyond Technical Fixes: Climate Solutions and the Great Derangement'. *Climate and Development* 12 (4): 343–352.

Noni, A. and K. Sanatomba. 2015. *Colonialism and Resistance: Society and State in Manipur.* New York: Routledge.

O'Dowd, M. F., and R. Heckenberg. 2020. 'Explainer: What Is Decolonisation?' *The Conversation*, 23 June. https://theconversation.com/explainer-what-is-decolonisation-131455 (accessed 24 December 2021).

Onta, N., and B. P. Resurreccion. 2011. 'The Role of Gender and Caste in Climate Adaptation Strategies in Nepal'. *Mountain Research and Development* 31 (4): 351–356.

OXFAM. 2020. 'Carbon Emissions of Richest 1 Percent More Than Double the Emissions of the Poorest Half of Humanity'. OXFAM International, 21 September. https://www.oxfam.org/en/press-releases/carbon-emissions-richest-1-percent-more-double-emissions-poorest-half-humanity (accessed 24 December 2021).

Pandey, K. 2021. '25 Years on, Many Indian States Haven't Implemented a Law That Empowers Adivasi Communities'. *Scroll.in*, 22 March. https://scroll.in/article/988729/25-years-on-many-indian-states-havent-implemented-the-law-that-empowers-adivasi-communities (accessed 24 December 2021).

Patnaik, S, and S. Jha. 2020. 'Caste, Class and Gender in Determining Access to Energy: A Critical Review of LPG Adoption in India'. *Energy Research and Social Science* 67. https://doi.org/10.1016/j.erss.2020.101530.

PIB Delhi. 2020. 'Coal India Limited to Produce 710 MT Coal in Current Financial Year: Shri Pralhad Joshi'. Press Information Bureau, Government of India, 23 April. https://pib.gov.in/PressReleasePage.aspx?PRID=1617419 (accessed 24 December 2021).

PNS. 2021. 'CCL Registers Highest Ever Coal Rake Dispatch on Single Day'. *The Pioneer*, 28 March. https://www.dailypioneer.com/2021/state-editions/ccl-registers-highest-ever-coal-rake--dispatch-on-single-day.html (accessed 24 December 2021).

Prasant, K., and D. Kapoor. 2010. 'Learning and Knowledge Production in Dalit Social Movements in Rural India'. In *Learning from the Ground Up: Global Perspectives on Social Movements and Knowledge Production*, edited by A. Choudry and D. Kapoor. New York, Palgrave, 193–210.

Rahi, A. 2019. 'AP PHOTOS: Indian Coal Mines Still Burning after a Century'. *Seattle Times*, 9 November. https://www.seattletimes.com/nation-world/nation/ap-photos-indian-coal-mines-still-burning-after-a-century/ (accessed 24 December 2021).

Ram, R. 2012. 'Reading Neoliberal Market Economy with Jawaharlal Nehru: Dalits and the Dilemma of Social Democracy in India'. *South Asian Survey* 19 (2): 221–241.

Ranganathan, M. 2022. 'Caste, Racialization, and the Making of Environmental Unfreedoms in Urban India'. Ethnic *and Racial Studies* 45 (2): 257–277.

Rao, A., and D. Kelleher. 2005. 'Is There Life after Gender Mainstreaming?' *Gender and Development* 13 (2): 57–69.

Rao, N., E. T. Lawson, W. N. Raditloaneng, D. Solomon, and M. N. Angula. 2019. 'Gendered Vulnerabilities to Climate Change: Insights from the Semi-arid Regions of Africa and Asia'. *Climate and Development* 11 (1): 14–26.

Rawat, R. S. 2011. *Reconsidering Untouchability: Chamars and Dalit History in North India.* Bloomington, IN: Indiana University Press.

Rawat, R. S. and K. Satyanarayana. 2016. *Dalit Studies.* Durham, NC: Duke University Press.

Reuters Staff. 2018. 'Why India Is Most at Risk from Climate Change'. World Economic Forum, 21 March. https://www.weforum.org/agenda/2018/03/india-most-vulnerable-country-to-climate-change (accessed 24 December 2021).

Ritchie, H. 2019. 'Who Has Contributed Most to Global CO2 Emissions?' *Our World in Data*, 1 October. https://ourworldindata.org/contributed-most-global-co2 (accessed 24 December 2021).

Routledge, P., A. Cumbers, and K. D. Derickson. 2018. 'States of Just Transition: Realising Climate Justice through and against the State'. *Geoforum* 88: 78–86.

Roy, A. 2020. 'The Pandemic Is a Portal'. *Financial Times*, 3 April. https://www.ft.com/content/10d8f5e8-74eb-11ea-95fe-fcd274e920ca (accessed 24 December 2021).

Roy, A., Benny Kuruvilla, and Ankit Bhardwaj. 2019. 'Energy and Climate Change: A Just Transition for Indian Labour.' *In India in a Warming World: Integrating Climate Change and Development,* edited by Dubash Navroz. New Delhi: Oxford University Press, 284–300.

Schlosberg, D., and L. B. Collins. 2014. 'From Environmental to Climate Justice: Climate Change and the Discourse of Environmental Justice'. *Wiley Interdisciplinary Reviews: Climate Change* 5: 359–374.

Sharma, M. 2012. 'Dalits and Indian Environmental Politics'. *Economic and Political Weekly* 47 (23): 46–52.

Shi, L., E. Chu, I. Anguelovski, A. Aylett, J. Debats, K. Goh, T. Schenk, K. C. Seto, D. Dodman, and D. Roberts. 2016. 'Roadmap towards Justice in Urban Climate Adaptation Research'. *Nature Climate Change* 6: 131.

Singh, R. K. 2021. 'Coal India Asks Government for Vaccines after Nearly 400 Covid Deaths'. *Bloomberg*, 8 June. https://www.bloomberg.com/news/articles/2021-06-08/top-coal-miner-seeks-more-vaccines-to-stem-india-covid-deaths (accessed 24 December 2021).

Skelton, A. et al. 2020. '10 Myths about Net Zero Targets and Carbon Offsetting, Busted'. *Climate Home News*, 11 December. https://www.climatechangenews.com/2020/12/11/10-myths-net-zero-targets-carbon-offsetting-busted/ (accessed 24 December 2021).

Sovacool, B. K., M. Burke, L. Baker, C. K. Kotikalapudi, and H. Wlokas. 2017. 'New Frontiers and Conceptual Frameworks for Energy Justice'. *Energy Policy* 105: 677–691.

Special Correspondent. 2021. 'Assam Solar Plant: Return Land, Withdraw Forces, Say Activists'. *The Hindu*, 12 May. https://www.thehindu.com/news/national/other-states/controversial-solar-plant-demand-to-withdraw-force-from-assam-farmers-land/article34540215.ece (accessed 24 December 2021).

Sreevatsan, A. 2018. 'British Raj Siphoned Out $45 Trillion from India: Utsa Patnaik'. *Livemint*, 21 November. https://www.livemint.com/Companies/HNZA71LNVNNVXQ1eaIKu6M/British-Raj-siphoned-out-45-trillion-from-India-Utsa-Patna.html (accessed 24 December 2021).

Staff Writer. 2021.' Delhi: Household Cooking Responsible for 40% of Pollution'. *Livemint*, 17 June. https://www.livemint.com/news/india/delhi-household-cooking-responsible-for-40-of-pollution-11623946808713.html (accessed 24 December 2021).

Suresh, R., J. James, and R. S. J. Balraju. 2020. 'Migrant Workers at Crossroads: The Covid-19 Pandemic and the Migrant Experience in India'. *Social Work in Public Health* 35 (7): 633–643.

Swarnakar, P. 2019. 'Climate Change, Civil Society, and Social Movement in India'. In *India in a Warming World: Integrating Climate Change and Development*, edited by N. K. Dubash. New Delhi: Oxford University Press, 253–272.

Tripathi, S. 2021. 'The Poem That's Channelling India's Anger about the Pandemic'. *The Guardian*, 28 May. https://www.theguardian.com/commentisfree/2021/may/28/poem-india-pandemic-gujarati-covid-narendra-modi (accessed 24 December 2021).

UN-DESA. 2018. '68% of the World Population Projected to Live in Urban Areas by 2050, Says UN'. United Nation Department of Economic and Social Affairs, 16 May. https://www.un.org/development/desa/en/news/population/2018-revision-of-world-urbanization-prospects.html (accessed 24 December 2021).

Varadhan, S. 2020. 'PM Modi Opens Coal Mining to Private Sector'. Reuters, 18 June. https://www.reuters.com/article/india-coal/pm-modi-opens-coal-mining-to-private-sector-idINKBN23P1KW (accessed 24 December 2021).

Vohra, K., A. Vodonos, J. Schwartz, E. A. Marais, M. P. Sulprizio, and L. J. Mickley. 2021. 'Global Mortality from Outdoor Fine Particle Pollution Generated by Fossil Fuel Combustion: Results from Geos-Chem'. *Environmental Research* 195 (April). https://doi.org/10.1016/j.envres.2021.110754.

Wu, J., D. Watkins, J. Williams, S. Venugopal Bhagat, H. Kumar, and J. Gettleman. 2020. 'Who Gets to Breathe Clean Air in New Delhi?' *New York Times*, 17 December. https://www.nytimes.com/interactive/2020/12/17/world/asia/india-pollution-inequality.html (accessed 24 December 2021).

Yenneti, K., and R. Day. 2015. 'Procedural (In)justice in the Implementation of Solar Energy: The Case of Charanaka Solar Park, Gujarat, India'. *Energy Policy* 86: 664–673.

ग्लोबल वॉर्मिंग

सुना है पृथ्वी पहले से ज़्यादा
गर्म हो रही है

धीरे धीरे यह गरमाहट
पृथ्वी के अक्ष को कुछ इस तरह
बदल रही है
कि

अपने अक्ष पर घूमने की
उसकी गति और तरीक़े
दोनों बदल सकते हैं

ऐसे में पृथ्वी का स्त्रीलिंग होना
क्या एक महज़ संयोग है ?

—संवेदना
१० अप्रैल '२०१६
मुंबई

Global Warming

The word is that the Earth
   is warming by the day
And this warming is
   gradually
changing the Earth's axis
so much so
   that it may change
its speed, its ways
Is it a mere coincidence
   that the Earth is thought of as a woman?
—*Translated by Veena Chhotray*

Samvedna Rawat's poem evokes a powerful sense of connection between women and the planet earth. By exploiting and degrading the planet's resources, we have debilitated planetary systems and thrust both the planet and marginalized groups into a crisis, not of their own making. However, the powerless—in this case, both planet earth and women—have a way of shaking things up. It is instructive that the poet does not use the frame of 'Mother Earth', which has often been used to paint an essentialized and apolitical understanding of planet Earth. Instead, the poem hints at the potential for healing rooted in the anger and power of the oppressed—just as Paulo Freire articulated in the *Pedagogy of the Oppressed*: 'It is only the oppressed who, by freeing themselves, can free their oppressors. The latter, as an oppressive class, can free neither others nor themselves.' The soul-melting heat of oppression is felt most intensely at the intersection of many cross-cutting identities and histories. So would the most potent paths toward healing and emancipation—of both the planet and its oppressed people.

CHAPTER 2

Urban Climate Justice in India

Eric Chu and Kavya Michael

Introduction

Indian cities are especially vulnerable to climate change due to their rapid population growth, high levels of socioeconomic inequality, and the general inability of infrastructure and public services to adapt to projected impacts (Revi 2008; Sharma and Tomar 2010). Although the neoliberal reforms introduced in India since the early 1990s have enabled the broader participation of non-state actors in decision-making, an ideological preference for entrepreneurial approaches to urban governance have largely led to the withdrawal of the state from delivering basic services (Datta 2015). Revenue shortfalls and lack of administrative capacity have further decreased the ability of cities to deal with climate impacts and risks (Cook and Chu 2018; Sharma et al. 2014). These effects are felt most acutely by the urban poor, who are disproportionately exposed (Michael and Vakulabharanam 2016; Satterthwaite et al. 2007).

Since the 1990s, there has been a growing awareness of climate change among government officials. For the next two decades, governmental interventions in Indian cities were confined to climate mitigation and targeted select manufacturing, construction, and energy sectors (Dubash et al. 2018). To be fair, climate adaptation was still a relatively nascent priority for India, and its policy focus was on furthering its geopolitical role in global climate negotiations. As a nation that saw itself as a rapidly industrializing global power, India aggressively pushed for the country's 'right to development' despite its significant exposure to climate change impacts (Gupta 2010). Indian negotiators highlighted how industrialized nations could support India through technology, resource, and capacity transfers that will allow it to 'leap frog' from fossil-fuel-intensive to more sustainable forms of development. Widespread awareness of climate adaptation only emerged in the late 2000s, spearheaded by transnational, civil society, and national scientific bodies that documented changing climatic patterns and advocated that subnational governments play a role in addressing climate risks (Khosla and Bhardwaj 2019b; Sharma, Singh, and Singh 2014; Sharma et al. 2014). Since then, and as climate adaptation has moved from the policy to the implementation space, there have been growing concerns that structural inequalities in urban development in India may dilute or even redirect the intended benefits of climate adaptation.

For cities across India, the combination of rapid urbanization and a changing climate has resulted in the disproportionate exposure of poor and marginalized communities to the impacts and associated risks of climate change (Chu and Michael 2019). The effects of climate change are mirrored in existing urban social relations of ethnicity, class, caste, gender, and other forms of power differentials, which are all arguably entrenched in forms of exclusion and inequality. For instance, Indian cities have, over the past several decades, transformed into spaces of wealthy enclaves and unplanned new towns at the periphery of older central cities (Vakulabharanam and Motiram 2012). Informal settlements at the urban periphery have precarious and insecure economies (Anand et al. 2014; Bhan and Jana 2015) where many residents are at risk of eviction due to insecure land tenure arrangements. Here, social structures characterized by marginalization and exclusion prevalent in rural villages are replicated (Shrivastava and Kothari 2012). Changing temperatures and precipitation levels, together with their cascading implications for health and housing, have only exacerbated such social inequalities.

In India, climate change policies – especially those concerning adaptation and resilience-building at the local scale – have often failed to recognize the particular needs of vulnerable sectors and communities. The urban poor, particularly the informal sector, often remain outside the ambit of urban planning mechanisms. Consequently, climate actions in Indian cities have remained exclusionary and have failed to address context-specific determinants of vulnerability and adaptive capacity (Chu and Michael 2019). In this chapter, we argue that theories of urban climate justice must go beyond including historically under-represented communities in decision-making and uncovering the distributive implications of climate, and must recognize intersecting and historically entrenched forms of socioeconomic, cultural, and political inequalities as well as the multiple channels through which climate change can exacerbate them.

Drawing on a longitudinal exploration of urban climate planning since the 1990s, this chapter assesses the structural drivers of climate injustice in Indian cities, with a focus on emerging adaptation and resilience priorities. We show examples from across the country of how drivers of injustice manifest in the design and documentation of adaptation actions as well as how they intersect to compound experiences of injustice. To further climate justice in Indian cities, we argue for a renewed focus on distributional, procedural, and recognitional justice from the bottom up. This may involve broadening civic dialogue around urban planning and practice to include demands for equity as the first step in reversing current exclusionary trends in urban development planning and climate policymaking. As a result, urban climate justice would be reoriented towards notions of inclusive development, human rights, and socioeconomic transformation.

Indian cities in a changing climate

Indian cities are increasingly facing the impact of climate change – temperature variability, droughts, flooding, cyclones, sea-level rise, and the linked environmental health risks – and are recognizing the need for climate adaptation and resilience-building. Poor communities are exposed to disproportionate risks from inadequate water, housing, sanitation, drainage, and solid waste management facilities. With its growing urban population, India will soon be one of the world's most vulnerable countries to climate change (Revi 2008; Yenneti et al. 2016). By the 2060s, it is expected that there will be approximately 500 million additional people living in an estimated 7,000 to 12,000 urban settlements across the country, most of whom will experience compounding environmental stressors relating to water, sanitation and environmental health, air and water pollution as well as climate change (Khosla and Bhardwaj 2019a; Sharma and Tomar 2010).

Historically, urban development was not a priority as the country relied heavily on the agricultural sector. However, the 74th Constitution Amendment Act (1992) provided formal recognition for urban local bodies and vested them with the power to undertake local sanitation, solid waste management, infrastructure, land provisioning, and development planning (Jayal, Prakash, and Sharma 2006). The Tenth and Eleventh Five Year Plans, designed for the years 2002–2012, both emphasized urban areas as engines of economic growth and advocated market-friendly reforms in urban infrastructure delivery. Under the Jawaharlal Nehru National Urban Renewal Mission (JNNURM), which ran from 2005 to 2014, public finances were directly allocated to cities. JNNURM adopted a governance reform-based funding approach, which meant that funds were supplied in conjunction with mandating reforms to local jurisdictional capacities and systems to enable urban infrastructure development and poverty alleviation across 65 cities (out of a total of 43,788 urban agglomerations and towns) (Khosla and Bhardwaj 2019a; Sharma and Singh 2016). A separate scheme, the Urban Infrastructure Development Scheme for Small and Medium Sized Towns (UIDSSMT), was launched in 2005 to support municipalities with smaller budgets and more capacity constraints (Sahasranaman 2012).

The central objective of these reforms was to decentralize larger (that is, Tier 1 and 2) cities as articulated under the 74th Constitution Amendment Act (1992) by strengthening public management and governance functions. Together with centrally-sponsored schemes such as Rajiv Awas Yojana, which ran from 2013 to 2014 and earmarked ₹322.3 billion for urban slum upgrading and poverty alleviation, the JNNURM served as an entry point to address questions of inadequate urban services delivery (Kundu 2014). Still, these schemes did not significantly address risk reduction, socioeconomic vulnerabilities, and climate adaptation to lower the overall impacts of increasingly extreme hazards. Also, although these reforms were not explicitly neoliberal (as opposed to those later articulated under the Smart Cities Mission), urban-level initiatives were often stymied by uncooperative state governments who were reluctant to transfer political, financial, or planning authority (Nandi and Gamkhar 2013).

During the same period – and spurred on by the approval of the National Action Plan on Climate Change (NAPCC) in 2008 – ministries at the national, state, and local levels began considering the implications of climate change for development functions. The NAPCC focused more on mitigation actions such as greenhouse gas reduction through reduced deforestation and regulation of industrial emissions and less on adaptation efforts. It also offered no financial provisions for climate action at the local level; hence, local governments continued to rely on intergovernmental disbursements schemes such as JNNURM. This approach was widely considered to be inadequate due to deficient capacities at the local level (Mehta and Mehta 2010).

Although there has never been an overt environmental agenda in urban planning in India, the confluence of ideas and opportunities presented by the policy mechanisms noted above began to spur actions to address climate change in cities. Some cities began to realize that infrastructure and service delivery investments must take into account climate impacts and support the local government's ability to address changing environmental risk profiles. These priorities have garnered increasing political traction in response to the escalating intensity of climate-related hazards. For example, three major cyclones – Helen (2013), Phalin (2013), and Hudhud (2014) – struck the Bay of Bengal coastal region within a short timeframe and Mumbai and Chennai both experienced devastating floods in 2015. Chennai also has a history of experiencing extreme heat (Jeganathan et al. 2016). These disasters laid bare the lack of preparedness and emergency planning and the fragility of the country's infrastructure.

In response, local governments availed of several intergovernmental schemes to support climate-resilient development, including the National Mission for Sustainable Habitat (2010), which emphasized building design, better urban planning, waste management, early warning systems, and regulatory reforms. Following the change in the central government in 2014, many of the schemes were revised to focus more on smart technologies and economic competitiveness in the context of sustainable development (Beermann et al. 2016; Fisher 2014). For example, the Atal Mission on Rejuvenation and Urban Transformation (AMRUT) was established in 2015 to channel ₹500 billion towards upgrading the urban water, transportation, and greenery sectors and the Swachh Bharat Abhiyan or Clean India Mission (2014–2019) promoted public health and sanitation across urban and rural areas. The flagship Smart Cities Mission, launched in 2016, budgeted nearly ₹980 billion (including matching funds from state governments) to support technological innovation in infrastructure and services provision. As of early 2021, 100 cities have been selected to receive funding primarily through area-based initiatives such as greenfield, transit, and service improvement projects.

Critiques of these schemes, particularly those enacted since 2014, have focused on their 'development first' approach, which has led to the side-lining of other priorities, particularly climate risk management and vulnerability reduction for the urban poor. The Smart Cities Mission has been explicitly critiqued for its neoliberal biases – for example, promoting special purpose vehicles to securitize debts for mega-infrastructure investors and developers and contracting out implementation efforts to private consulting and engineering firms. Further, though more than 5,000 projects were proposed, there remains some level of uncertainty regarding actual disbursements, expenses incurred over time, and the proportion of budgetary allocations that were actually spent on implementing smart projects. In other cases, large urban development projects were favoured, as they enabled the creation of world-class elite cities. This political shift corresponded with a global surge in resilience thinking (Bohland, Davoudi, and Lawrence 2019), which promoted the idea that local governments should be resistant to a wide array of political, economic, and environmental shocks (Borie et al. 2019).

However, in India and across the Global South, resilience thinking has been criticized for its focus on technocratic solutions and a tendency to overlook historically entrenched socioeconomic inequalities. At the same time, a reliance on public–private partnerships and speculative land investments has increased economic inequality and social exclusion (Bahadur and Thornton 2015; Chu 2020).

For example, a green housing project in the outskirts of Bengaluru named Towards Zero Carbon Development (T-Zed) promotes low carbon living by effectively combining green forms of consumption with urban development (Bulkeley and Castán Broto 2014). However, this project has little impact on ongoing inequalities within the city, especially when more than 35 per cent of the population lives in poor informal settlements that are highly vulnerable to climate impacts (Kumar, Geneletti, and Nagendra 2016). Instead, the project channels resources towards creating a gated community for a growing market of high-earning, green-minded middle-class residents.

The myriad policy advancements in India over the past 30 years mostly support the greater involvement of the private sector in urban development and a withdrawal of the state from delivering basic services (Goldman 2011; Vakulabharanam and Motiram 2012). This has led to land speculation and acquisition of land for special economic zones, dispossession of the working class through slum evictions, prioritization of private sector interests, and the emergence of new parastatal bodies, special purpose vehicles, and quasi-autonomous bodies to govern cities (Chattopadhyay 2017). Climate action also follows this logic, leading to a surge in middle-class environmentalism that largely ignores the structural causes of climate vulnerabilities and risks (Chu and Michael 2019). The experience of climate injustice, therefore, stems from the interaction between historically entrenched socioeconomic inequalities and development constraints that can be attributed to recent neoliberal governance reforms, superimposed on a reality of increasingly severe climate change impacts.

Emerging focus on climate adaptation and resilience

Awareness of climate adaptation as something separate from disaster risk reduction was introduced in India by multilateral aid and philanthropic actors in the late 2000s. India had a robust regulatory framework for addressing disaster impacts, which drew from its experiences managing extreme events such as Cyclone Phailin in 1999 and the Kutch earthquake in 2001 (Jha, Basu, and Basu 2016; Pal, Ghosh, and Ghosh 2017). This framework was eventually codified through national and state disaster management agencies. Prioritization of climate adaptation policies targeting long-term climate stressors such as heat, precipitation, and sea-level rise took longer. Low awareness was compounded by the uneven implementation of the 74th Constitution Amendment Act (1992), which led to the unclear division of planning and governance responsibilities across urban, state, and national authorities. Local institutional complexities further stymied climate adaptation efforts as policy responsibilities were disaggregated across urban bureaucratic functions (through the municipal corporation) and land use management and planning functions (through the urban development authority).

For many cities, climate adaptation priorities were also driven by external capacities, resources, and policy support. Significant effort was needed to localize climate models to arrive at projections of heat, precipitation, and sea-level change, especially since such technical capacities did not typically exist within local governments. International organizations such as the German Agency for International Cooperation (GIZ), United Nations Development Programme (UNDP), World Bank, the Rockefeller Foundation, ICLEI-Local Governments for Sustainability, and, to a lesser extent, the US Agency for International Development (USAID), helped introduce climate adaptation ideas and language in local planning and policymaking in India. There were also several bilateral partnerships between donors and local governments – for example, Kolkata's partnership with UK Aid, which was formalized in 2013. These initiatives initially focused on understanding how changing rainfall, temperature, flooding, and sea-level rise would affect infrastructure and urban communities. As awareness was low, they focused on assessing which productive sectors were most exposed to climate impacts as well as which sections of society were most vulnerable to climate risks.

Early programmes, such as those helmed by the Rockefeller Foundation's Asian Cities Climate Change Resilience Network (ACCCRN) – with their pilot efforts in Indore, Surat, and Gorakhpur – prioritized the integration of climate science into planning, management, and governance mechanisms through relatively representative processes. A focus on procedural representation was prioritized given the high levels of uncertainty and lack of understanding of the degree to which economic and social sectors were exposed to different heat, precipitation, and flooding impacts. Creating participatory arenas aided in co-generating locally relevant information on socioeconomic vulnerabilities in hotspots of concentrated risk such as flood plains, riverine settlements, and informal communities. Representative processes were generally commended for successfully uncovering the key vulnerabilities and risks facing cities, while structured participatory methodologies such as 'shared learning dialogues' facilitated discussions on common problems among previously disparate urban leaders and bureaucrats (Sharma and Singh 2016). As such, early advances in cross-sectoral communication and problem-solving within cities were identified as key innovations.

However, researchers have retrospectively critiqued these early advances by asserting that historically marginalized and vulnerable communities continued to be excluded from formal planning processes, which subsequently led to negative outcomes for them (Anguelovski et al. 2016; Shi et al. 2016). For example, although plans from Kota, Rajasthan, identified slum populations as especially vulnerable, the subsequent decision-making and planning processes did not meaningfully engage representatives from this group (Wilk et al. 2018). Rockefeller-led efforts prioritized identifying empathetic city leaders to help improve awareness of climate impacts, assess urban vulnerability, and identify projects that could both highlight the benefits of proactive adaptation actions and potential ways to integrate them with ongoing development priorities (Brown 2018). Given the relative lack of awareness, a conscious coupling (or mainstreaming) of climate adaptation with on-the-ground basic services, housing, health, and economic development priorities made political sense. Although this approach took time and effort, it allowed adaptation priorities to gain a foothold in cities and helped channel financial resources and coordinate project designs.

Between 2008 and 2014, the Rockefeller Foundation and ICLEI-Local Governments for Sustainability attempted to scale up adaptation action to other cities using a less resource-intensive approach. This meant less handholding, a condensed assessment process, and a more structured approach to drafting local resilience strategies. By 2014, several additional cities produced resilience strategies, including Kochi, Visakhapatnam, Bhubaneswar, Shimla, Mysore, Nainital, Patna, and Gangtok, but the degree to which the recommendations were implemented by the local administration is unclear. The scaled-up phase was less successful, as cities had less incentive to participate and the condensed time frame made climate adaptation resemble an externally driven development project rather than a genuine internal programme with local buy-in, resource support, and leadership. Several cities, such as Kochi and Visakhapatnam, showed some evidence that climate priorities had been integrated into city disaster management plans and city development plans with provisions to engage civil society organizations in first response and security actions during disaster events. But other externally led initiatives suffered as long-term institutionalization of climate priorities in urban planning, development, and governance was met with resistance.

By 2014, political and ideological changes in the national government led to widespread changes in how climate change priorities were articulated at the policy level. The mantra of urban resilience rather than climate adaptation or climate risk management gained a foothold through various government schemes that consolidated economic progress, human security, and, to a lesser extent, environmental sustainability under one large banner. A new wave of intervention targeted the creation of smart and resilient cities – exemplified by the Smart Cities Mission (2015) – but simultaneously placed renewed financial constraints on local governments through the enactment of the Good and Services Tax (GST), which replaced previous intergovernmental disbursement mechanisms such as the JNNURM. Under the new tax regime, local governments were no longer guaranteed revenue as state governments were not obliged to disburse it to them (in fact, many did not). Domestic policy changes also mirrored changes in global institutional priorities, with the Rockefeller Foundation launching the 100 Resilient Cities (100RC) initiative around the same time.

Evidence from the field

Early urban climate adaptation plans across India helped identify policy champions and relevant resources to further the nascent agenda, although these efforts were later found to generally exclude perspectives from historically disadvantaged groups. For example, even in a relatively rich city like Mumbai, research has shown that differences in wealth and capacity account for high levels of household vulnerability (Romero-Lankao, Gnatz, and Sperling 2016). Early plans were critiqued for providing a surface-level acknowledgement of the different socioeconomic vulnerabilities faced by the urban poor while failing to address structural drivers of inequality and unequal exposure to risks. These drivers of vulnerability can be attributed to the neoliberal political reforms introduced since the early 1990s, which have led to the broad privatization of urban services, unequal distribution of economic opportunities, and increasing concentration of political authority among elites (Joshi 2014).

In Table 2.1, we explore recent climate adaptation and resilient development plans across 19 Indian cities, ranging from small to large and inland to coastal municipalities. Our intention is not to offer a comprehensive or exhaustive survey of climate adaptation and resilience actions; instead, Table 2.1 provides a snapshot of experiences and approaches to either strategically or comprehensively operationalize climate priorities within existing land use, infrastructure, risk management, or wider planning processes. We include standalone adaptation and resilience plans, sector-specific policies (such as those targeting urban heat impacts), and more general disaster management and sustainability strategies that prioritize climate adaptation. Our goal is to offer a quick view of select efforts on the ground, drawing on the authors' own research and policy engagements in various cities, while also highlighting the different actors, interests, and resource pathways involved in the process. We build upon ongoing comparative efforts (see Khosla and Bhardwaj 2019b; Singh et al. 2021) by offering insights on how to identify climate injustices on the ground and shed light on approaches that can enable more just and equitable adaptation actions going forward.

Table 2.1 Analysis of key social equity or justice dimensions in recent urban climate change plans in India

City	Plan	Consideration of Justice	Key Approaches to Promoting Equity/Justice
Ahmedabad, Gujarat	Heat Action Plan (2017)	High	The plan identified populations that are vulnerable to extreme heat during the summer months. The municipality was charged with creating a list of high-risk areas for extreme heat and organizing preventative training and outreach efforts for local communities. Actions included expanding cooling centres and shaded areas for outdoor workers, slum communities, and migrants.
Bhubaneswar, Odisha	City Disaster Management Plan* (2014)	Low	The plan acknowledged that several urban sectors and communities are more vulnerable to disaster impacts (heat waves, floods, earthquakes, fires, epidemics, and so on). It integrated community-level actions, including local risk and vulnerability assessments and training programmes in schools.

(Contd)

(*Contd*)

City	Plan	Consideration of Justice	Key Approaches to Promoting Equity/Justice
Chennai, Tamil Nadu	Chennai City Resilience Strategy (2019)	High	The plan exhibited an understanding of the compounding and structural nature of vulnerabilities. Resilience-building was linked with social security stability and justice. The plan recognized that protecting vulnerable communities was a key pillar for building the city's resilience. However, it emphasized upgrading informal settlements, which has led to questions of unaffordability.
Delhi, NCT	Climate Change Agenda for Delhi 2009–2012 (2009)	Low	The plan focused on technical and engineering solutions such as solar energy, air pollution mitigation, building and construction standards, energy efficiency, water resources use and distribution, and urban greening. It had minimal engagement with questions of socioeconomic inequality and vulnerability.
Gorakhpur, Uttar Pradesh	Towards a Resilient Gorakhpur (2010)	Medium	The plan recognized the lower adaptive capacities of urban poor communities. Interventions focused on social advocacy in diverse communities as well as knowledge and awareness campaigns.

(*Contd*)

(*Contd*)

City	Plan	Consideration of Justice	Key Approaches to Promoting Equity/Justice
Guwahati, Assam	Climate Proofing Guwahati: City Resilience Strategy and Mainstreaming Plan (2013)	Low	The plan highlighted the lack of planning and housing provisions in slum areas leading to higher vulnerability (especially to floods). It noted that poor or sub-standard infrastructure services increase the vulnerability of the population to disasters and climate-related extreme events.
Indore, Madhya Pradesh	City Resilience Strategy for Changing Climate Scenarios (2012)	Medium	The plan recognized that migrants and informal settlements are particularly vulnerable to climate impacts. Strategies focused on housing, sewage, drainage, water access, and other services for the urban poor.
Jorhat, Assam	Climate-Ready City: Strategy for Building Resilience to Urban Climate Change (2017)	Low	The plan recognized the climate vulnerabilities of underserved low-income communities, especially in terms of health, housing, and access to medical services. It also identified some community-based adaptation strategies.
Kochi, Kerala	Development Plan for Kochi City Region 2031* (2020)	Low	The plan took a mitigation approach to climate change. Even though the document referred to inclusive development and delivery of basic urban services, the link with climate-induced disasters was not fully articulated.

(*Contd*)

(*Contd*)

City	Plan	Consideration of Justice	Key Approaches to Promoting Equity/Justice
Kolkata, West Bengal	Roadmap for Low Carbon and Climate Resilient Kolkata (2016)	Low	The strategy highlighted climate change's intersections with public health, air pollution, urban heat, water, green spaces, solid waste management, and transportation priorities. It indicated differential vulnerabilities across the city.
Panaji, Goa	Revised City Development Plan for Panaji* (2015)	Low	The plan focused on ecological impacts and key risks to infrastructure and identified low-income areas that are vulnerable to floods and water inundation. The plan included sections on urban poor and low-income communities. Some adaptation options focused on 'social infrastructure' but there is no specific mention of social equity.
Pune, Maharashtra	Pune Resilience Strategy (2019)	Medium	The plan acknowledged the need for equitable and inclusive growth, particularly for migrant labourers and low-income groups. It included provisions to support access to affordable housing and civic participation in planning. It focused on social cohesion and inclusivity (in the context of stability, security, and justice) rather than directly mentioning inequality, but spoke of informal economic opportunities and poverty reduction.

(*Contd*)

(Contd)

City	Plan	Consideration of Justice	Key Approaches to Promoting Equity/Justice
Rajkot, Gujarat	Heat Wave Action Plan (2018)	Low	The plan was based on an assessment of vulnerable areas and communities. Strategies included knowledge dissemination in slum communities.
Saharsa, Bihar	City Resilience Strategy: Sahara City (2017)	Low	The plan recognized the disproportionate vulnerability of informal and migrant settlements. Adaptation strategies focused on information and awareness-building among community members, as well as strategies to improve housing, infrastructure, and service provision.
Shimla, Himachal Pradesh	Climate Resilient Strategy: Shimla City (2013)	Low	The plan acknowledged the vulnerability of certain populations and sectors, including informal settlements, street vendors, women, and tourists, but it failed to mention inclusive planning processes.
Surat, Gujarat	Surat Resilience Strategy (2017)	Medium	The plan recognized the differential vulnerability of the poor to flooding, heat, and public health risks. It focused on affordable housing, mobility, social cohesion, and health service provision for the poor. It included strategies for inclusive decision-making, primarily stakeholder workshops.

(Contd)

(*Contd*)

City	Plan	Consideration of Justice	Key Approaches to Promoting Equity/Justice
Thiruvananthapuram, Kerala	City Disaster Management Plan* (2015)	Low	The plan identified some socioeconomic vulnerabilities and the need for community-level strategies (such as community centres) in response to disaster impacts.
Visakhapatnam, Andhra Pradesh	City Disaster Management Plan* (2013)	Low	The plan focused on disaster response and relief mechanisms, although it did identify vulnerable urban areas and communities along the coast and in low-lying areas. It advocated for long-term resilience, with some focus on the well-being of vulnerable localities, children, and public health concerns. It highlighted the role of non-governmental organizations (NGOs) as volunteers and first responders, particularly during extreme heat events.

Source: Authors' synthesis.

Note: * denotes analysis focused on the climate change sections of a larger plan.

A high-level overview shows that some cities, such as Pune and Chennai, have produced city-wide resilience strategies with funding support from the 100RC programme. The programme provided member cities with funding for instituting a salaried Chief Resilience Officer position within a high-level municipal department as well as resources to support comprehensive planning efforts. Kolkata similarly benefited from UK development aid for drafting a combined mitigation and resilience strategy. Other cities, such as Jorhat and Saharsa, built upon the legacy of civil society support – in this case, the Gorakhpur Environmental Action Group (GEAG) – to enable community-based approaches to resilience planning. Still other cities such as Bhubaneswar, Visakhapatnam, and Thiruvananthapuram elected to integrate emerging climate adaptation priorities into ongoing city disaster management plans, which had been mandated by their respective state governments given their high exposure to natural disasters. Finally, cities such as Ahmedabad and Rajkot focused on one climate impact – urban heat – and devised specific strategies to respond to it.

Most of the plans highlighted in Table 2.1 were drafted between 2009 and 2019 and apply external expertise to translate scientific models into urban social and economic scenarios. The climate projections drew upon data from national scientific agencies such as the Indian Meteorological Department and the National Disaster Management Authority and research organizations such as The Energy and Resources Institute. The areas of planning focus varied according to local contextual needs, ranging from disaster risk management, urban heat, and flooding to general urban economic transitions in the context of climate change. For example, some cities noted the role of technology and infrastructure in response to climate impacts, such as in the Roadmap for Low Carbon and Climate Resilient Kolkata (2016). Many plans recognized the differential forms of vulnerability experienced by low-income, informal, and migrant communities, such as the higher levels of exposure to heat, flooding, and disaster impacts. For example, Ahmedabad's Heat Action Plan (2017) noted the need for more cooling centres and shaded areas catering to outdoor workers and slum and migrant communities. Indore's City Resilience Strategy for Changing Climate Scenarios (2012), Guwahati's City Resilience Strategy and Mainstreaming Plan (2013), and Panaji's City Development Plan (2015) all acknowledged that informal communities are more vulnerable to flooding, inundation, and subsequent health risks. Visakhapatnam's City Disaster Management Plan (2013) and Shimla's Climate Resilience Strategy (2013) further showed how women, children, and the elderly are additionally vulnerable.

Beyond differential vulnerability, several cities explicitly targeted procedural equity concerns by recognizing the need to include community voices in decision-making. Some plans drew on inclusive and participatory planning processes, engaging with community leaders and civil society organizations to design and evaluate plans and policies. For example, the Surat Resilience Strategy (2017) and Towards a Resilient Gorakhpur (2010) detailed participatory efforts that, to various extents, included local government officials, community leaders, and NGOs in the planning process. Research has shown that these efforts are key to ensuring the legitimacy of decision-making processes, although questions remain around whether such arrangements are truly representative of diverse interests and include the voices of disadvantaged groups (Chu 2016b, 2020). A second strategy for including community voices is harnessing community-based adaptation strategies. For instance, Bhubaneswar's City Disaster Management Plan (2014) and Jorhat's Strategy for Building Resilience to Urban Climate Change (2017) advocated for community disaster response teams, local water provisioning systems, as well as community-led mobilization to support resource and capacity distribution in the event of disasters. These strategies drew on the recommendations articulated by numerous state-level disaster management authorities to develop volunteer and civil defence groups to respond to natural disasters.

The examples highlighted in Table 2.1 indicate uneven progress in tackling social equity and justice priorities in ongoing urban climate actions. In addition to not having shared criteria for assessing equity and justice, many cities, in fact, rely on NGOs and external funders to sustain baseline participatory processes. Among the 19 cities highlighted in Table 2.1, we see two broad approaches to climate equity and justice: recognizing differential vulnerability and including community-based adaptation and response strategies. It is important to acknowledge the reality that climate risks are unequally distributed among communities and that exposure to impacts depends on the quality of shelter, employment security, and access to crucial water, education, transport, and energy services. However, as we argue in this chapter, this view of equity only considers immediate, near-term access to goods and capacities but does not fully address the underlying drivers of poverty, vulnerability, and marginality. Furthermore, many plans do not articulate efforts to include previously unrepresented voices in the design and evaluation of strategies. Cities often rely on preexisting strong social networks while ignoring others or rely on locally dominant public–private or civil society partnerships at the expense of minority interests.

Towards urban climate justice

Insights from Indian cities suggest that emerging climate efforts, especially those that do not rely on NGOs or external funder support, rarely go beyond surface-level participatory practices to redress structural factors and processes that make the urban poor vulnerable to climate change. Plans tend to focus on instruments, strategies, and actions required to rectify immediate distributive inequalities rather than diagnose the structural factors contributing to social, economic, and political marginality. This section situates evidence from Indian cities within broader urban climate justice scholarship and highlights potential strategies to enable justice and equity going forward. More specifically, we note that to promote more radical and progressive visions of climate justice, planning processes in Indian cities must better consider four dimensions of climate justice: (1) addressing the differential distribution of climate impacts among the urban poor, (2) tackling the root causes of climate vulnerability, (3) delineating shared responsibilities for inclusive decision-making, and (4) pursuing intersectional forms of climate justice. We briefly elaborate on these four dimensions below.

First, a pivot towards justice requires us to recognize that urban poor communities are differentially exposed to the impacts and risks of climate change. Our chapter has shown that climate impacts exert additional stressors on already vulnerable urban communities and compound experiences of socio-political domination, infrastructure exclusion, and economic exploitation. Climate hazards can cause loss of land and livelihoods, putting pressure on the city's existing infrastructure (Michael, Deshpande and Ziervogel 2019; Revi 2008). Furthermore, climate impacts are often unequally distributed due to inadequate poverty alleviation programmes, social exclusion, lack of investment in public services and infrastructure, and gaps in skill, capacity, and knowledge development. For instance, a vast majority of India's informal workers reside in precarious locations across cities and their peripheries. The vulnerability of informal workers is compounded by insecure housing tenure rights and lack of employment opportunities and access to basic services (Anand et al. 2014; Bhan and Jana 2015). Social divisions and hierarchies based on caste and gender further accentuate experiences of poverty. Thus, urban climate actions must first seek to redress differential forms of exposure and vulnerability on the ground.

Second, there is a need to tackle the root causes of climate vulnerability and the legacy of unequal development in cities. As we have highlighted earlier, there is evidence that climate vulnerability and marginality have been exacerbated by governance reforms enacted in India in the past few decades. Reforms since the mid-2010s have promoted entrepreneurial and extractive approaches to urban development, as evidenced by numerous intergovernmental schemes that privilege public–private partnerships and the financialization of infrastructure and services (Datta 2015; Desai and Sanyal 2012). Local governments are therefore incentivized to generate revenue through financially speculative – and often exploitative – means, thereby side-lining priorities such as public welfare, social support, and poverty alleviation. In India, even without considering climate change, forms of urban marginalization are the outcomes of historic development pathways that have yielded highly unequal processes and patterns of allocating resources and access to spaces within the city (Shrivastava and Kothari 2012; Vakulabharanam 2010). This has further resulted in benefits for a particular socioeconomic class and uneven power relations across society (Chattopadhyay 2017). Efforts to realize climate justice on the ground must therefore tackle these longstanding trends in development inequality, exclusion, and dispossession.

Third, there is a need to delineate shared responsibilities with respect to inclusive climate change decision-making and action in cities. In this chapter, we have noted that there has been a gradual veering towards more technical interventions that draw on top-down schemes, external funds, and public–private implementation mechanisms (Chu 2016a; Khosla and Bhardwaj 2019b). Examples of this include the emerging role of transnational organizations, parastatal agencies, and top-down initiatives driven by central directives or external development projects, often focused on environmental actions that benefit the elite or upper-middle class. Therefore, adaptation and resilience actions are constrained by a lack of autonomy, limited resources, low awareness, low bureaucratic stability, the siloed nature of climate actions, and a disconnect between technical climate knowledge and embodied experiences of environmental risks. Despite these complexities, however, some cities have managed to carve out more participatory arenas that have helped translate external climate knowledge into local development priorities. A shared language has emerged around the need to address climate impacts and risks, and new forms of civil society networks have been established to support more inclusive local decision-making. For instance, several examples highlighted in Table 2.1 involve strategies to enact far-reaching adaptation programmes by uncovering co-benefits between climate adaptation, mitigation, and livelihoods protection or by including local, community-based action. Still, as highlighted already, most of these actions are yet to tackle the structural drivers of development inequality that gave rise to unequal exposure to climate impacts and risks in the first place.

Finally, there is a need to pursue intersectional considerations of climate justice that span social groups. An intersectional approach to climate justice seeks to articulate forms of structural inequality based on gender, class, caste, race/ethnicity, and citizenship status (Chu and Cannon 2021; Matin, Forrester, and Ensor 2018; Rao et al. 2019; Wilson and Chu 2020). For example, the informal economy in Indian cities is largely constituted by excluded masses that subsidize and feed the formal economy by providing various cheap inputs in the form of labour or commodities. There is evidence that the needs of women, migrants, and informal communities are often not taken into account in existing climate adaptation and resilience plans (Chu and Michael 2019; Michael, Deshpande, and Ziervogel 2019). The growing importance of unpaid female labour further solidifies traditional gender norms. It exists to support the survival of male migrants in hostile urban conditions – care activities and the provision of basic needs like cooking, cleaning, and fetching water is allocated to women (Rao 2017). From a climate justice point of view, groups that are intersectionally marginalized, such as women in the informal economy, are likely to have fewer opportunities to influence policymaking, so decisions made by governments are unlikely to benefit them. As such, a pivot towards intersectionality in climate justice will help illuminate the differential experiences of vulnerability of different social groups due to their position in power structures and context-specific, dynamic social categories (Cannon and Chu 2021).

Conclusion

Indian cities are emblematic sites of environmental and developmental inequality, featuring spatial concentrations of poverty and informality. There is emerging literature on climate mitigation and adaptation at the sub-national level in India, but most of it concerns how sub-national entities are responding to global and national goals in terms of parameters such as carbon emissions, financing, and infrastructure provision (Dubash et al. 2018). There has not been a strong focus on lived experiences, developmental dilemmas, and embodied forms of inequality within cities (Khosla and Bhardwaj 2019b). Thus, Indian cities need to rethink their approach to climate action through the lens of justice. By surveying the historical trajectory of how Indian cities have addressed key climate risks and vulnerabilities, this chapter has demonstrated how the maldistribution of climate impacts must be understood in light of development deficits linked to the country's neoliberal economic transformation over the past three decades. As highly unequal spaces, cities house burgeoning informal settlements with concentrated socioeconomic and environmental vulnerabilities, where socio-cultural divisions around gender, class, caste, and religion are exacerbated (Sultana 2014). Projected climate impacts such as flooding, sea-level rise, droughts, and health crises exert additional stressors on an already unequal development context.

Emerging theories on climate injustice in Indian cities must consider the structural disenfranchisement experienced by the poor. Urban climate justice should place equal emphasis on distributive, procedural, and recognition equity to tackle the drivers of climate inequality (Chu and Michael 2019). We, therefore, call on climate change scholars and activists to envision more radical approaches to tackling the differential drivers of climate vulnerability and the root causes of development inequality, while also pursuing more inclusive decision-making processes and devising intersectional strategies to effect climate justice on the ground. Despite these aspirations, enacting such a radical reorientation in climate action in Indian cities will be challenging. Evidence shows that local plans are increasingly socially exclusive; climate actions still reflect logics tied to financial bankability, and multilateral actors are continuing to rely on speculative forms of infrastructure and service provision. In response, just and equitable forms of climate action in Indian cities must go beyond addressing the maldistribution of climate-induced losses and benefits to furthering the recognition of minority voices and redressing the highly unequal distribution of human capabilities and developmental rights.

References

Anand, S., G. Bhan, C. Idicheria, A. Jana, and J. Koduganti. 2014. *Locating the Debate: Poverty and Vulnerability in Urban India*. IIHS-RF Paper on Urban Poverty. Bangalore, India: IIHS.

Anguelovski, I., L. Shi, E. Chu, and H. Teicher. 2016. 'Equity Impacts of Urban Land Use Planning for Climate Adaptation: Critical Perspectives from the Global North and South'. *Journal of Planning Education and Research* 36 (3): 333–348.

Bahadur, A. V. and H. Thornton. 2015. 'Analysing Urban Resilience: A Reality Check for a Fledgling Canon'. *International Journal of Urban Sustainable Development* 7 (2): 196–212.

Beermann, J., A. Damodaran, K. Jörgensen, and M. A. Schreurs. 2016. 'Climate Action in Indian Cities: An Emerging New Research Area'. *Journal of Integrative Environmental Sciences* 13 (1): 55–66.

Bhan, G., and A. Jana. 2015. 'Reading Spatial Inequality in Urban India'. *Economic and Political Weekly* 50(22): 49–54.

Bohland, J., S. Davoudi, and J. L. Lawrence (eds.). 2019. *The Resilience Machine*. New York and London: Routledge.

Borie, M., M. Pelling, G. Ziervogel, G., and K. Hyams. 2019. 'Mapping Narratives of Urban Resilience in the Global South'. *Global Environmental Change* 54: 203–213.

Brown, A. 2018. 'Visionaries, Translators, and Navigators: Facilitating Institutions as Critical Enablers of Urban Climate Change Resilience'. In *Climate Change in Cities: Innovations in Multi-Level Governance*, edited by S. Hughes, E.K. Chu, and S.G. Mason. Cham: Springer International Publishing, 229–253.

Bulkeley, H. and C. Castán Broto. 2014. 'Urban Experiments and Climate Change: Securing Zero Carbon Development in Bangalore'. *Contemporary Social Science* 9 (4): 393–414.

Cannon, C. E. B. and Chu, E. K. 2021. 'Gender, Sexuality, and Feminist Critiques in Energy Research: A Review and Call for Transversal Thinking'. *Energy Research and Social Science* 75: 102005.

Chattopadhyay, S. 2017. 'Neoliberal Urban Transformations in Indian Cities: Paradoxes and Predicaments'. *Progress in Development Studies* 17 (4): 307–321.

Chu, E. 2016a. 'The Governance of Climate Change Adaptation through Urban Policy Experiments'. *Environmental Policy and Governance* 26 (6): 439–451.

———. 2016b. 'The Political Economy of Urban Climate Adaptation and Development Planning in Surat, India'. *Environment and Planning C: Government and Policy* 34 (2): 281–298.

———. 2020. 'Urban Resilience and the Politics of Development'. In *Climate Urbanism: Towards a Critical Research Agenda*, edited by V. Castan Broto, E. Robin, and A. While. Cham: Springer Publishing, 117–136.

Chu, E. K., and C.E. Cannon. 2021. 'Equity, Inclusion, and Justice as Criteria for Decision-Making on Climate Adaptation in Cities'. *Current Opinion in Environmental Sustainability* 51: 85–94.

Chu, E., and Michael, K. 2019. 'Recognition in Urban Climate Justice: Marginality and Exclusion of Migrants in Indian Cities'. *Environment and Urbanization* 31 (1): 139–156.

Cook, M. J., and E. K. Chu. 2018. 'Between Policies, Programs, and Projects: How Local Actors Steer Domestic Urban Climate Adaptation Finance in India'. In *Climate Change in Cities: Innovations in Multi-Level Governance*, edited by S. Hughes, E. Chu, and S. Mason. Cham: Springer, 255–277.

Datta, A. 2015. 'New Urban Utopias of Postcolonial India: "Entrepreneurial Urbanization" in Dholera Smart City, Gujarat'. *Dialogues in Human Geography* 5 (1): 3–22.

Desai, R., and R. Sanyal (eds.). 2012. *Urbanizing Citizenship: Contested Spaces in Indian Cities.* New Delhi, India: Sage Publications.

Dubash, N. K., R. Khosla, U. Kelkar, and S. Lele. 2018. 'India and Climate Change: Evolving Ideas and Increasing Policy Engagement'. *Annual Review of Environment and Resources* 43 (1): 395–424.

Fisher, S. 2014. 'Exploring Nascent Climate Policies in Indian Cities: A Role for Policy Mobilities?' *International Journal of Urban Sustainable Development* 6 (2): 154–173.

Goldman, M. 2011. 'Speculative Urbanism and the Making of the Next World City'. *International Journal of Urban and Regional Research* 35 (3): 555–581.

Gupta, J. 2010. 'A History of International Climate Change Policy'. *Wiley Interdisciplinary Reviews: Climate Change* 1 (5): 636–653.

Jayal, N. G., A. Prakash, and P. Sharma (eds.). 2006. *Local Governance in India: Decentralization and Beyond.* Oxford and New York: Oxford University Press.

Jeganathan, A., R. Andimuthu, R. Prasannavenkatesh, and D. S. Kumar. 2016. 'Spatial Variation of Temperature and Indicative of the Urban Heat Island in Chennai Metropolitan Area, India'. *Theoretical and Applied Climatology* 123 (1–2): 83–95.

Jha, A., R. Basu, and A. Basu. 2016.' Studying Policy Changes in Disaster Management in India: A Tale of Two Cyclones'. *Disaster Medicine and Public Health Preparedness* 10 (1): 42–46.

Joshi, S. 2014. 'Environmental Justice Discourses in Indian Climate Politics'. *GeoJournal* 79 (6): 677–691.

Khosla, R., and A. Bhardwaj. 2019a. 'Urban India and Climate Change'. In *India in a Warming World*, edited by N. K. Dubash. Oxford and New York: Oxford University Press, 459–476.

———. 2019b. 'Urbanization in the Time of Climate Change: Examining the Response of Indian Cities'. *Wiley Interdisciplinary Reviews: Climate Change* 10 (1): e560.

Kumar, P., D. Geneletti, and H. Nagendra. 2016. 'Spatial Assessment of Climate Change Vulnerability at City Scale: A Study in Bangalore, India'. *Land Use Policy* 58: 514–532.

Kundu, D. 2014. 'Urban Development Programmes in India: A Critique of JnNURM'. *Social Change* 44 (4): 615–632.

Matin, N., J. Forrester, and J. Ensor. 2018. 'What Is Equitable Resilience?' *World Development* 109: 197–205.

Mehta, M. and D. Mehta. 2010. 'A Glass Half Full? Urban Development (1990s to 2010)'. *Economic and Political Weekly* 45 (28): 20–23.

Michael, K., T. Deshpande, and G. Ziervogel. 2019. 'Examining Vulnerability in a Dynamic Urban Setting: The Case of Bangalore's Interstate Migrant Waste Pickers'. *Climate and Development* 11 (8): 667–678.

Michael, K., and V. Vakulabharanam. 2016. 'Class and Climate Change in Post-reform India'. *Climate and Development* 8 (3): 224–233.

Nandi, S. and S. Gamkhar. 2013. 'Urban Challenges in India: A Review of Recent Policy Measures'. *Habitat International* 39: 55–61.

Pal, I., T. Ghosh, and C. Ghosh. 2017.' Institutional Framework and Administrative Systems for Effective Disaster Risk Governance: Perspectives of 2013 Cyclone Phailin in India'. *International Journal of Disaster Risk Reduction* 21 (January): 350–359.

Rao, N., A. Mishra, A. Prakash, C. Singh, A. Qaisrani, P. Poonacha, et al. 2019. 'A Qualitative Comparative Analysis of Women's Agency and Adaptive Capacity in Climate Change Hotspots in Asia and Africa'. *Nature Climate Change* 9 (12): 964–971.

Rao, S. 2017. 'Women and the Urban Economy in India: Insights from the Data on Migration'. In *Gender and Time Use in a Global Context: The Economics of Employment and Unpaid Labor*, edited by R. Connelly and E. Kongar. New York: Palgrave Macmillan US, 231–257.

Revi, A. 2008. 'Climate Change Risk: An Adaptation and Mitigation Agenda for Indian Cities'. *Environment and Urbanization* 20 (1): 207–229.

Romero-Lankao, P., D. M. Gnatz, and J. B. Sperling. 2016. 'Examining Urban Inequality and Vulnerability to Enhance Resilience: Insights from Mumbai, India'. *Climatic Change* 139 (3–4): 351–365.

Sahasranaman, A. 2012. 'Financing the Development of Small and Medium Cities'. *Economic and Political Weekly* 47 (24): 59–66.

Satterthwaite, D., S. Huq, H. Reid, M. Pelling and P. Romero-Lankao. 2007. *Adapting to Climate Change in Urban Areas: The Possibilities and Constraints in Low- and Middle-Income Nations*. London, UK: IIED.

Sharma, D. and S. Tomar. 2010. 'Mainstreaming Climate Change Adaptation in Indian Cities'. *Environment and Urbanization* 22 (2): 451–465.

Sharma, D., R. Singh, and R. Singh. 2014. 'Building Urban Climate Resilience: Learning from the ACCCRN Experience in India'. *International Journal of Urban Sustainable Development* 6 (2): 133–153.

Sharma, D. and S. Singh. 2016. 'Instituting Environmental Sustainability and Climate Resilience into the Governance Process: Exploring the Potential of New Urban Development Schemes in India'. *International Area Studies Review* 19 (1): 90–103.

Sharma, V., V. Orindi, C. Hesse, J. Pattison, and S. Anderson. 2014. 'Supporting Local Climate Adaptation Planning and Implementation through Local Governance and Decentralised Finance Provision'. *Development in Practice* 24 (4): 579–590.

Shi, L., E. Chu, I. Anguelovski, A. Aylett, J. Debats, K. Goh, et al. 2016. 'Roadmap towards Justice in Urban Climate Adaptation Research'. *Nature Climate Change* 6 (2): 131–137.

Shrivastava, A. and A. Kothari. 2012. *Churning the Earth: The Making of Global India*. New York: Penguin.

Singh, C., M. Madhavan, J. Arvind, and A. Bazaz. 2021. 'Climate Change Adaptation in Indian Cities: A Review of Existing Actions and Spaces for Triple Wins'. *Urban Climate* 36. https://doi.org/10.1016/j.uclim.2021.100783.

Sultana, F. 2014. 'Gendering Climate Change: Geographical Insights'. *Professional Geographer* 66 (3): 372–381.

Vakulabharanam, V. 2010. 'Does Class Matter? Class Structure and Worsening Inequality in India'. *Economic and Political Weekly* 45 (29): 67–76.

Vakulabharanam, V. and S. Motiram 2012. 'Understanding Poverty and Inequality in Urban India since Reforms: Bringing Quantitative and Qualitative Approaches Together'. *Economic and Political Weekly* 47 (47/48): 44–52.

Wilk, J., A. C. Jonsson, , B. Rydhagen, A. Rani, and A. Kumar. 2018. 'The Perspectives of the
     Urban Poor in Climate Vulnerability Assessments: The Case of Kota, India'. *Urban Climate*
     24 (June 2017): 633–642.
Wilson, J. and E. Chu. 2020. 'The Embodied Politics of Climate Change: Analysing the Gendered
     Division of Environmental Labour in the UK'. *Environmental Politics* 29 (6): 1085–1104.
Yenneti, K., S. Tripathi, Y. D. Wei, W. Chen, and G. Joshi. 2016. 'The Truly Disadvantaged?
     Assessing Social Vulnerability to Climate Change in Urban India'. *Habitat International* 56:
     124–135.

Let Justice Bloom by Anupriya

CHAPTER 3

How Just and Democratic Is India's Solar Energy Transition?

An Analysis of State Solar Policies in India

Karnamadakala Rahul Sharma and Parth Bhatia

Introduction

In our warming world, energy provision is not simply about technology but also politics (Hughes and Lipscy 2013). Energy systems are the result of intensely contested political battles in the domains of technology selection, ownership of capital, environmental externalities, access, and siting. The geographical reach, terms of access, and forms of ownership of electricity infrastructures reflect the prevailing distribution of political and economic power (Bridge, Özkaynak and Turhan 2018). Consequently, this gives rise to injustices such as uneven electricity access, displacement, and voicelessness among marginalized communities. Control over energy infrastructure is not just the result but often also the source of political and social power (Amin 2014; Larkin 2013) – that is, energy shapes politics just as much as politics shape energy.

India is facing the twin imperatives of tackling historic energy poverty through an expansion of its energy system on the one hand and pursuing climate mitigation on the other. India's electricity sector is dominated by coal-fired thermal power, which in turn drives the country's carbon emissions. The energy sector as a whole contributed around 74 per cent of India's total greenhouse gas (GHG) emissions in 2015, of which 38 per cent was from public electricity generation (GPI Secretariat 2016). On the other hand, India's average monthly residential electricity consumption is only

90 kilowatt-hour (kWh), which is one-third of the global average and one-tenth of that of the US (Chunekar and Sreenivas 2019). Despite official estimates of 100 per cent electrification, many households still receive poor quality electricity for only a few hours each day (S. D'Souza 2019). The growing feasibility of renewable energy (RE) indicates a potential opportunity to address both climate mitigation and energy poverty challenges. India announced a target of 450 gigawatt (GW) of RE by 2030 as against a total installed capacity of 370 GW in April 2020 (PMO India 2019). As we progress towards a low-carbon system, what are the implications of this transition, given existing patterns of injustice and the prospects of their reproduction in our twenty-first-century energy infrastructure?

India's electricity system can be characterized by its gigantic scale; the primary state ownership of its generation, transmission, and distribution infrastructure; cross-subsidization from commercial and industrial consumers to agricultural consumers; and its federal nature. Due to the unique technical characteristics of solar photovoltaics (PV) – modularity, intermittency, and fuel-free generation – it offers an opportunity to fundamentally disrupt the political, financial, and institutional arrangements associated with the existing system[1] (Dubash, Swain and Bhatia 2019; Stephens 2019). These potential disruptions include attracting high-paying industrial consumers away from the grid, allowing new players (individuals, co-operatives, high-risk fast capital) to compete for energy ownership, and shifting the federal balance of power as the Centre's monopoly over coal loses salience.

By disrupting the existing equilibrium of power, the rise of renewables offers an opportunity to link energy choices to broader social justice goals and to redistribute power and wealth within societies (Angel 2016; Stephens 2019). Whether the ultimate beneficiary of an RE-based society is the common energy user instead of the elite will be contingent on how new energy infrastructures are specifically structured and will not be simply determined by the choice of technology. It will hinge upon whether the RE-based system incorporates the concerns of the marginalized, compensates the losers of this transition (such as coal workers), shares benefits inclusively, and creates participatory forms of governance. This is where the critical lenses of energy democracy and energy justice gain salience in India.

This chapter explores the extent to which India's state-level solar energy policies embody the goals of a democratic and just energy transition. We first define and

---

[1] Modularity is a feature of PV technology, which means that the constituent unit is small in scale, but many such modules can be combined to create a system of any size. In contrast, conventional power systems have many sub-components and only become economically feasible at large scales.

contrast energy justice and democracy. Second, we examine how these conceptual lenses have been applied in India and present a framework for analysis. Third, we explain our methodology and then discuss findings from an analysis of key state solar policies. Fourth, we conclude by contextualizing the insights from the energy system by locating it within the broader theme of climate justice and by offering avenues for further research in this field. Before delving further, it is worth highlighting that our chapter focuses primarily on the policy discourse surrounding practices of RE deployment and does not engage directly with the growing literature on the impact of energy transitions on the coal sector. Our focus on RE serves to complement the coal-focused chapter in this volume by Vasudha Chhotray (Chapter 6).

Understanding energy democracy and energy justice

Energy justice is a conceptual agenda that aims to evaluate 'where injustices emerge, which affected sections of society are ignored and which processes exist for their remediation in order to reveal, and reduce such injustices' (Jenkins et al. 2016, 175). The literature on energy justice provides the conceptual and analytical guidance needed to assess and resolve energy-related dilemmas, both in terms of outcomes and procedures (Sovacool and Dworkin 2015).

The three main constituent elements of energy justice are procedural, distributive, and recognition justice (McCauley et al. 2013). A fourth tenet, restorative justice, has also been proposed by scholars as a way to repair the harm done to people (and/ or society/nature) in the past (Heffron and McCauley 2017). Another important framework of energy justice is the eight-principle decision-making framework, which provides tools for policymakers to operationalize energy justice in policy frameworks (Sovacool and Dworkin 2015). The focus has historically been on incorporating procedural and distributional justice into policy frameworks, while recognition concerns have received more limited attention.

The concept of energy democracy emerged at a trade union roundtable organized by the Global Labour Institute at Cornell University in 2012 (Stephens 2019). Energy democracy was framed in terms of three objectives: *resist* the agenda of fossil fuels corporations, *reclaim* to the public sphere parts of the energy economy that have been privatized or marketized, and *restructure* the global energy system to massively scale up RE and other safe low-carbon options, implement energy conservation, and ensure job creation and true sustainability (Sweeney 2012). Burke and Stephens (2017, 35) defined it as 'an emergent social movement advancing RE transitions by resisting the fossil-fuel-dominant energy agenda while reclaiming and democratically restructuring energy regimes'. Szulecki (2018, 35) defines it as a quasi-utopian 'political goal, in which citizens are the recipients, stakeholders and accountholders of the entire energy sector policy'. While there are disparate conceptualizations of energy democracy, one of the core demands of this movement is for publicly owned and democratically managed energy systems (Burke 2018).

In contrast, discussions in the Global South have historically centred on questions of energy access, energy poverty, institutional distortions (corruption), and enhancing recognition of the needs of marginalized communities, including women (Guruswamy 2011; Lacey-Barnacle, Robison, and Foulds 2020). Of the two, energy justice has found greater resonance in the Global South, whereas energy democracy is still primarily centred in the Global North (Lacey-Barnacle, Robison, and Foulds 2020). We speculate that countries that have a tradition of civic engagement in utility management are more likely to provide fertile ground for energy democracy ideas to take root. For instance, rural electricity distribution in most of the United States is organized through consumer-owned rural electric co-operatives (RECs), over 800 of which continue to deliver ~11 per cent of the total units of electricity sold in the US (University of Wisconsin Center for Cooperatives 2020).

More broadly, the discursive and political context in India is fundamentally different from that of the West (Angel 2016). Here, the justice conversation is dominated by the challenge of access, which is not a major concern in developed countries (Malakar, Herington, and Sharma 2019). Moreover, these discussions assume that the energy system is controlled by a democratic state that presumably supports decentralized RE as part of its developmental discourse. Such assumptions are rarely borne out in the varied contexts of the developing world.

The Indian context

In India, electricity is largely generated using conventional sources of energy such as coal, large hydropower, gas, and nuclear, and a fraction comes from utility-scale solar and wind. Consumers largely play a passive role in this system – they receive electricity, pay a recurring bill, and have limited avenues to participate in electricity planning. Where participation does exist, it usually concerns land acquisition and is often very limited in scope. Decision-making and implementation are carried out by central and state regulators; the ministries dealing with power, coal, and RE; large, corporatized utilities (state-owned or private); grid operators; and frontline staff engaged in billing and maintenance. There are not many avenues for consumers to exercise their voice beyond inefficient consumer grievance channels and sparsely attended public hearings.

The thrust of the RE policy is driven by factors such as energy security, attracting private investment, and domestic political signalling (Shidore and Busby 2019). A vast majority of RE capacity is privately owned as opposed to conventional sources, due to the general push towards privatization in the energy sector since the 2000s (Moallemi et al. 2017). In this sense, the broader public has lesser control over India's RE capacity base than it has over the thermal capacity base, which largely involves public sector undertakings (PSUs). Given this institutional context, the transition to a democratic energy system might seem unlikely.

Nevertheless, energy democracy has entered the discourse on energy transitions in recent years in India, and most notably from the labour movement. Mathews, Barria, and Roy (2016, 2), writing under the banner of Trade Unions for Energy Democracy, lay out a 'core labour perspective' for a just energy transition. The two key political battlegrounds identified by them include the lessening of labour's bargaining power due to an RE policy that favours the private sector and securing democratic rights for communities being displaced by large-scale solar parks. Their vision of energy democracy includes four key demands: (a) rehabilitation of coal areas, (b) redeployment and retraining of the coal sector workforce, (c) ensuring financing for the transition, and (d) public-sector-led and municipalities-controlled RE development. Further, they call for participatory spaces 'where mass organisations and trade unions democratically engage and shape industrial policy' (Mathews, Barria, and Roy 2016, 13).

As a complement to the focus on the coal sector articulated earlier, our chapter explores the opportunities for democratic transitions using RE. While doing so, it is important to critically assess the normative value attached to all forms of RE and not unequivocally equate RE penetration with advancing energy justice and democracy in the Global South. First, justice effects are not inherent to the expansion of RE, but depend on choices concerning scale, siting, and ownership of RE (Banerjee et al. 2017). Second, a normative preference for renewables over traditional sources like biomass and charcoal has been characterized as an 'elitist interpretation of modernist development ideology' resulting from the lack of a nuanced understanding of traditional sources (Munro, van der Horst, and Healy 2017, 640). Third, the Global North has been accused of 'energy bullying' or promoting RE development that would benefit corporations based there (Monyei et al. 2018, 2019; Todd et al. 2019).

In sum, energy justice and energy democracy are powerful tools for any country in the process of finalizing its energy trajectory, but they need to be applied carefully in the context of developing countries. This includes adapting key frameworks to suit the local context as we seek to do in the following section.

Analytical approach and research methods

While energy justice and democracy have their own unique histories, they are interrelated, and policy instruments that contribute to one can reinforce the other. Policies play an important role in establishing the direction of change and the rules of the game. To the best of our knowledge, no other work to date has examined energy justice and democracy from the perspective of state policies in India. We analyse both concepts in this study through the identification of policy clauses that move us towards fair distribution and more democratic procedures.

Energy is a concurrent subject in the Indian federal system. While the Centre sets the overall trajectory through planning and financing, it is at the state level that policy implementation and distribution of electricity occurs. The financial support provided by the Centre and its priorities provide structure to the overall electricity system's transition. However, states control important levers that influence the realization of distributional and procedural goals. States also vary in their approach to governance (such as the extent of decentralization), which can influence and inform their approach to electricity governance (Dubash, Kale, and Bharvirkar 2018). Identifying the creative ways in which some states have accommodated justice concerns within the federal framework demonstrates the feasibility of achieving a more just policy framework.

Our analysis is rooted in this context and reads these policies using an Indian and, more broadly, Southern lens. Our reading of distributional justice begins with the question of access since this continues to be a dominant challenge in the Indian context. Cross-sectoral initiatives that distribute the benefits of electrification through employment and increased economic activity are central to our interpretation of distributional justice. With respect to more democratic procedures, our analysis accounts for the low purchasing power among domestic consumers of electricity and the low levels of financial and personnel capacity required for decentralized management. The elements of our framework are detailed in the following paragraphs.

First, with respect to access (distributional goals), we look for both the *identification* of underserved groups and the *recognition* of their specific electrification needs. Identification occurs when a target group is mentioned in the policy as a potential beneficiary of better electricity access. Recognition pushes the conversation beyond connectivity and asks whether these underserved communities can afford and use electricity over the long term. For example, electrification of poor rural households needs to recognize their particular spending patterns and prior experience with metering systems. KWh-based metering and monthly billing cycles impose informational and financial burdens on the poor (Winkler et al. 2011). On the other hand, service-based charges or fixed daily payments mimic existing expenditure patterns on energy services in poor households (Sharma, Palit, and Ramakrishnan 2016). Due to their modularity and zero fuel cost, distributed and decentralized solar energy offer more opportunities than the legacy system for restructuring business models to ease access for the poor. Several alternative models of service provisioning have been attempted by the private and non-profit sectors and can offer guidance on how to transition poor consumers to cleaner sources of electricity (Bhattacharyya 2013). In our review of state policies, we identify specific instances where state policies on RE move from simple identification to the recognition of consumers' needs.

Second, a truly distributive system must aim to not just redistribute electricity access but the developmental benefits accruing from electricity. The strong relationship between energy and development is well established in the literature (Alstone, Gershenson, and Kammen 2015). However, small quantities of electricity supplied at the household level do little to improve socioeconomic outcomes (Aklin et al. 2017). In addition, rural enterprises require several important non-electricity inputs to achieve growth and financial sustainability (Ganguly et al. 2020, Willcox et al. 2015). Policies seeking to distribute the benefits of electrification more fairly, therefore, need to do more than just focus on electricity supply and should actively seek cross-sectoral integrations. This would require coordination and integration across multiple domains such as skilling, human resource development, enterprise development, and education. However, such overarching strategies and goals for integration are often not supported by instruments that increase coordination and convergence through the provisioning of governing resources such as funds, legislative orders, and interdepartmental working groups or the explicit integration with existing government programmes (Candel and Biesbroek 2016; Candel 2019). In our analysis, we explore the strength of and variation in coordination mechanisms across states.

Third and finally, we discuss democratic procedures that allow for wider, more inclusive, and fair public participation in RE deployment. Based on a review of the literature, we identify three sets of instruments to meet procedural goals: instruments that (a) facilitate ownership and ease transactions, (b) decentralize legacy institutions, and (c) enhance just participation. The primary goal of instruments that facilitate ownership and ease transactions is to increase the amount of RE used by consumers or fed into the grid. Net-metering policies link individuals and the grid by allowing users to consume as well as sell the electricity generated by their solar power systems. These instruments aim to create *prosumers* – individuals or groups that both produce and consume energy. In certain business models, ownership is transferred to new market participants who can either save money on utility bills or earn an income by selling excess electricity to the grid. While existing literature considers net and gross metering to be 'key policies for energy democracy', they might only mark an incremental step towards democratization in certain contexts (Burke and Stephens 2017, 39). For example, if the rights and responsibilities of rooftop owners, tariffs, and regulations are all strongly controlled by central and state regulators, these instruments can end up only facilitating exchange or transactions but not ownership. In our analysis, we seek to highlight policies that go beyond just offering metering options and create opportunities for more consumer participation.

Our second set of instruments includes those that decentre legacy institutions and pave the way for decentralized institutions to manage electricity. Co-operatives, farmers associations, and self-help groups are commonly recognized as institutions of decentralized governance in the extant literature and policy discourse. These institutions can facilitate a transition towards a more democratic energy system because they are already built on the idea of community participation. However, there are three reasons why they might not aid a democratic transition at scale. First, managing complex infrastructure such as electricity will require a significant amount of capacity-building. Experience with rural electrification projects involving village energy committees (VECs) or voluntary groups constituted for the management of decentralized solar energy systems has been mixed (Chaurey et al. 2012; Palit et al. 2013). VECs often do not have the manpower or technical capacity for managing local energy systems and need support from technical partners over the long term (Sharma et al. 2014; Sharma and Palit 2020). Second, these groups are largely membership-driven organizations without an electoral mandate. This raises questions about their representation and their accountability toward the larger community. There is also the possibility of elite capture, which makes them an ineffective partner in the transition to more democratic systems. Third, these groups do not have the same status as the government departments that they have to engage with during the implementation and management of decentralized electricity systems. They are likely to face significant hurdles in transacting with the government machinery given their unequal share of power in governance processes. This is where existing, *elected* institutions of decentralized governance such as village panchayats and urban local bodies (ULBs) are likely to be better candidates for facilitating a democratic electricity transition. While short-term implementation goals might be achieved by transferring ownership to community groups, a longer-term vision for democratic transition must consider the involvement of elected institutions of governance. In our analysis, we will identify cases where policies have looked beyond voluntary groups and associations, and have sought to empower elected institutions of governance by involving them in decision-making regarding energy production and management.

The third set of instruments promotes just participation. In the Indian context, displacement and loss of livelihoods resulting from infrastructure development are well documented. Development-induced displacement has been studied in the case of the Sardar Sarovar Dam on the Narmada river, the displacement of residents of urban informal settlements during the Delhi Commonwealth Games, and in mining, among other sectors (Baviskar 1995; Kohli 2013). On similar lines, Yenneti and Day (2015) offer in-depth case studies of the lack of procedural justice in the Charanka Solar Park project in Gujarat, which led to the displacement of local communities and loss of livelihoods. In this study, we seek to identify instruments that foster a participatory approach that considers the livelihoods of local communities in solar energy transitions.

Finally, Indian scholarship on RE has highlighted the emancipatory potential of decentralized electricity systems (R. D'Souza 2019). In 1960, D. D. Kosambi argued for decentralized solar energy managed by communities without any aid from the government. For him, this was the only form of technology that would realize a truly socialist energy system 'without the stifling effects of bureaucracy and heavy initial investment' (R. D'Souza 2019, 42). Amulya Reddy was another influential advocate for democratizing energy who advocated for the self-reliance of villages through employment-generating, community-owned, off-grid energy systems. This vision has shaped the RE debate in India for many years, until recently. Since the Electricity Act 2003, the thrust of electricity policy has been towards liberalizing electricity generation, adding capacity primarily through large thermal powerplants, and expanding grid-based access. The final section of our analysis gives a big picture view of the current status of centralized and decentralized electricity systems and India's progress towards a just and democratic electricity system.

Methods

We analysed the latest versions of the notified solar energy policies of each state, as uploaded on the RE departments' websites. There is wide variation in the formats of these policies – some states have a single document, while others have two or three different documents for large-scale solar power plants (grid-connected, utility-scale solar power) and decentralized solar power (also referred to as distributed generation, mini-grids, or decentralized distributed generation) or rooftop systems. Further, some states have an RE policy covering multiple sources and no individual solar energy policies, while some have both. If the RE policy was the only available policy, we only reviewed the solar energy section within it. If both RE and solar policies exist, we reviewed only the solar policy in cases where the RE policy was ratified earlier. Some states also have solar-hybrid policies and, where available, these have been included for review. Any amendments to the latest version of the solar or RE policies have also been included.

Our analysis involved a close reading of the policy documents to identify from the preamble, objectives, and clauses the references made to the *distributional and procedural goals* of the proposed solar energy transition. Clauses within policies that mention such goals were manually highlighted and coded into a worksheet along with the clause and page numbers. We then examined the occurrence and objectives of such clauses across policies by employing the lens of interpretive policy analysis (Yanow 2007).

Data and observations

The distribution of different types of policy documents across states is depicted in Table 3.1.

Analysis and discussion

In terms of distributional goals, we find that most policies continue to exclude significant marginalized groups. Where groups are included, the focus is more on identifying them rather than recognizing their specific needs or the processes by which they can effectively transition to becoming full consumers of electricity. We also noticed that distributional goals beyond simple access are mentioned in the preambles of policies, but are not substantiated by an allocation of tools to foster the cross-sectoral collaboration required for their implementation.

Among the three sets of procedural instruments, those that facilitate ownership and ease transactions were emphasized and elaborated on more than those that decentre legacy institutions or enhance just participation. With the exception of invoking urban municipal bodies to amend by-laws to facilitate rooftop solar, the limited attention given to new institutional arrangements and just processes of participation reflects the norms of the legacy electricity system. Overall, the policies tend to keep the system in its current configuration and forego the opportunity solar provides to create transformative change beyond reducing emissions.

In the next three subsections on distributional and procedural goals, we have used examples from different states to elucidate the various sets of instruments for operationalizing energy justice in Indian solar policies. We have also conducted

Table 3.1 Source of solar power policies across states

State	Single State Solar Policy	RE Policy	Multiple Policies	Hybrid Policies	Amendments
Andhra Pradesh	■			■	■
Arunachal Pradesh					
Assam	■				
Bihar		■			
Chhattisgarh	■				
Goa	■				
Gujarat	■			■	■
Haryana	■	■			■
Himachal Pradesh	■				
Jharkhand	■		■		
Karnataka	■				■
Kerala	■	■			
Madhya Pradesh	■	■			
Maharashtra	■				■
Manipur	■				
Meghalaya		■			
Mizoram	■	■			
Nagaland					
Odisha	■	■			
Punjab	■	■			■
Rajasthan	■			■	■
Sikkim	■				
Tamil Nadu	■				
Telangana	■				
Tripura	■				
Uttar Pradesh	■		■		
Uttarakhand	■				■
West Bengal		■			

(Contd)

(*Contd*)

State	Single State Solar Policy	RE Policy	Multiple Policies	Hybrid Policies	Amendments
J&K	■	■	■	■	■
Delhi	■				

Note: The black boxes represent the policy documents included in the analysis. Two states (Maharashtra and Chhattisgarh) had policy documents only available in regional languages and have thus been excluded from our analysis.

a comprehensive assessment of state policies, using the methods described in the earlier section titled 'Analytical approach and research methods', to identify the presence of policy instruments for achieving energy justice goals. These results are synthesized in Table 3.2.

Distributional goals: identification versus recognition

All the reviewed solar policies include provisions for greater distribution. The most commonly identified target groups are farmers and residents of remote and rural areas who are not connected to the national electricity grid. Policies suggest standalone solar pumps for farmers and either standalone solar home systems or community-level mini-grids for remote and rural locations.

This is a straightforward concern about distribution – farmers and remote communities are indeed important groups from a distribution perspective. However, the policies fail to mention women, residents of urban informal settlements, and nomadic and pastoral groups. There is evidence of gender-based disparity in electricity access and use and lack of access in urban informal settlements and among pastoralist groups (Baruah 2015; Debnath et al. 2020). Given the nature of their electricity demands, mobility, and low-paying capacities, these groups can be particularly well-served by decentralized and small-scale solar power.

The reviewed policies also fall short of recognizing the specific needs of the populations they wish to serve through solar energy transitions, with two exceptions. The Kerala 2013 Solar Energy Policy takes a step towards recognition by stating that 'for consumers with monthly consumption of 30 units and below efforts shall be made involving welfare departments of Government and LSGIs (Local Self Government Institutions) to solar enable them and in such cases, a special feed-in-tariff scheme shall be notified' (Government of Kerala 2013, 7). While there is insufficient information to draw any conclusions about outcomes, in terms of intent, this provision suggests that consumers with very low loads of below 30 units a month need to be given special tariff considerations. The Karnataka Solar Policy 2014–2021

Table 3.2 Presence of policy instruments for energy justice in state solar policy documents

	Distributional goals		Procedural goals		
	Recognition beyond identification	Cross-sectoral integration for justice	Facilitation of ownership and easing transition	Decentering of legacy institutions	Enhancement of just participation
Andhra Pradesh			*		
Arunachal Pradesh	████	████	████	████	████
Assam	*			**	
Bihar	*	**		**	
Chhattisgarh	████	████	████	████	████
Goa					
Gujarat	*	**			
Haryana	*			*	
Himachal Pradesh	*		**		**
Jharkhand	*	*	**	**	
Karnataka	**	**	*	**	
Kerala	**			**	**
Madhya Pradesh	*				
Maharashtra		*			
Manipur					
Meghalaya					
Mizoram	*	**	*		*
Nagaland	████	████	████	████	████
Odisha	*	*		**	
Punjab			**		
Rajasthan	*	**	*	**	**
Sikkim	*	**	**	*	
Tamil Nadu			**	**	*
Telangana				**	**
Tripura	████	████	████	████	████
Uttar Pradesh	*	**			

(Contd)

(*Contd*)

	Distributional goals		Procedural goals		
	Recognition beyond identification	Cross-sectoral integration for justice	Facilitation of ownership and easing transition	Decentering of legacy institutions	Enhancement of just participation
Uttarakhand					
West Bengal	*			**	**
Jammu & Kashmir					**
Delhi		**	**	**	

Note: ** (double asterisks) represents a clear or strong occurrence and * (single asterisk) represents a partial or weak presence of policy instruments in the relevant category. Grey cells imply that the policy or translation was unavailable. The empty cells represent '*gaps*' or a lack of any instruments for the category.

provides exceptional financial assistance of ₹1 crore for small solar parks (but >100 acre in size) located in 'backward districts' (Government of Karnataka 2014, 10). Similar to Kerala, there is insufficient information on whether the needs of these districts are recognized beyond mentioning that the solar parks must be small.

Our findings bring up the question of whether we should expect policy documents to go into such detail; after all, they are meant to offer broad guidance. Here, we point to the discrepancy in the extent of detail provided for policy clauses relevant to underserved populations and those relevant to wealthier urban residents or corporations. Most policies focus on promoting new business models and strategies to increase the penetration of utility-scale and rooftop solar power plants, none of which embody distributional goals. These include multiple business models for solar rooftop power plants, detailed net and gross metering policies, bidding guidelines, and land acquisition procedures, among other enabling policy mechanisms. A more holistic vision of a just transition needs to look beyond replicating the metrics of the legacy electricity system and move towards recognizing the specific needs, spending patterns, information asymmetries, and transaction costs associated with different target groups in accessing electricity.

Distributional goals: cross-sectoral integration for justice beyond access

Across all policies, the preamble and objectives emphasize *(a)* transitioning the electricity system towards cleaner sources of energy, *(b)* energy security, and *(c)* serving marginalized populations. Several policies, however, aim to extend the scope of their goals beyond the electricity sector and mention sustainable development, jobs, and creating rural enterprises. This second set of policy goals are fundamentally distributive in nature, as they seek to provide the developmental benefits accruing from electrification to previously underserved populations. Referring to our framework, however, we find little evidence that such goals are supported by instruments to enable cross-sectoral collaboration, with a few exceptions.

Some state policies refer to a mechanism for training and absorbing unemployed youth into the solar industry mentioned in India's national solar energy programme – the Jawaharlal Nehru National Solar Mission. While it has been mentioned in the policies, there is no indication of policy integration at the state level. Two state policies stand out in terms of seeking explicit convergence with non-electricity sector policies that could lead to employment generation. Bihar's Renewable Energy Policy of 2017 aims to forge partnerships for skill development and capacity-building with the existing Bihar Rural Livelihoods Project, JEEViKA, to 'reach out to local youth especially women to support entrepreneurship at the grass-root level, to improve socioeconomic conditions of financially underprivileged' (Government of Bihar 2017, 16). On similar lines, Gujarat's Solar Policy from 2015 (Government of Gujarat 2015) explicitly makes linkages to existing industrial development programmes to enable convergence, specifically with the Gujarat Industrial Policy of 2015 and the Electronics Policy for the State of Gujarat (2014–2019), both of which extend state-level incentives for the development of RE and semiconductors (Government of Gujarat 2015, 21).[2] Besides programme convergence, creating institutional structures to coordinate cross-sectoral activities is also important. A few states, such as Delhi, Karnataka, Rajasthan, and Mizoram, have constituted empowered committees consisting of officials from departments such as power, urban development, PWD, environment, and finance, typically under the chairmanship of the chief secretary. In summary, while some exceptions exist, there are limited instruments across states to enable the much-needed cross-sectoral collaboration for meeting broader distributional goals.

---

[2] Several states have provisions to ensure convergence between building codes and solar energy use. West Bengal proposes mandatory installation of Solar PV rooftop systems. Other policies such as those from Delhi, Rajasthan, Odisha, Sikkim, and Jharkhand also propose reframing building codes for facilitating solar energy installations. We mention this as a footnote since this convergence, while important, does not directly address our point on framing convergence as a means to achieve greater justice.

Procedural goals

Instruments that facilitate ownership and ease transactions

All the reviewed policies focus extensively on instruments such as net and gross metering. The term *prosumer* is used across policies in sections describing solar rooftop systems and a range of business models to support uptake is described. However, as we argue in our analytical framework, while such instruments play a role in re-distribution, they might only make an incremental shift towards more democratic ownership.

A few policies imagine metering beyond facilitating transactions. The policies of Delhi, Jharkhand, and Sikkim include virtual metering in addition to net and gross metering (Government of Jharkhand 2018, 4; Government of NCT of Delhi 2016, 7; Government of Sikkim 2019, 6). Virtual net metering allows potential prosumers without rooftops to invest in community rooftop systems, either within their neighbourhoods or outside them. While the exchange of electricity with the grid remains the same as with net metering, this policy innovation deepens participation in two ways: first, consumers who would otherwise be unable to install a rooftop system now can. This would be particularly relevant in dense urban areas. Second, this can, in turn, increase the size of the community investing in decentralized systems, leading to a greater potential for bargaining power.

Instruments that decentre legacy institutions

Co-operatives, farmers' associations, and self-help groups are commonly recognized as new institutions of decentralized governance. However, as argued earlier, their limited capacity to manage complex infrastructure, non-representativeness, and lower status compared to government departments limit the scope of their contribution in the transition to a democratic energy future at scale. Instead, elected institutions of decentralized governance must also be considered. Some policies move us in this direction by indicating that panchayats and municipalities can play a role in managing and implementing solar power plants. Bihar's Renewable Energy Policy 2017, for example, notes the role of 'registered companies, government entities, partnership companies/firms, individuals, consortia, *Panchayat Raj Institutions, Urban Local Bodies*, Co-operative or registered society [*sic*]' (Government of Bihar 2017, 3). Kerala's Solar Energy Policy 2013 similarly emphasizes the role of local self-governments in power production and proposes introducing 'incentive[s] for people's representatives/panchayats [to promote] solar installations and street light optimization', making a rare reference to representative government entities

(Government of Kerala 2013, 6). The West Bengal RE Policy 2012 explicitly states that 'urban local bodies will form an essential part of the comprehensive solar policy for cities' (Government of West Bengal 2012, 17). Some states like Assam and Jharkhand go a step further by proposing the amendment of municipal by-laws to facilitate the adoption of solar rooftop systems (Government of Assam 2018; Government of Jharkhand 2018).

Instruments that enhance just participation

In 2017, the Ministry of New and Renewable Energy relaxed the requirements for environmental and social impact assessments (EIA/SIA) for utility-scale solar power projects, including solar parks. This is reflected in the state solar policies released subsequently. Some state policies, however, do take steps to ensure fair compensation for communities whose land is being acquired for solar energy projects. Himachal Pradesh's policy states that '1% of the total cost of the project, as fixed by HPERC (Himachal Pradesh Electricity Regulatory Commission)', will be paid to the Local Area Development Fund for 'community development works', for government land on which people have community rights (Government of Himachal Pradesh 2016, 12). Similarly, Telangana's policy states that 'development charges and layout fee of INR 25,000 per acre basis shall be levied payable to the respective Panchayat', in the section on 'Ease of Business: Enabling Provisions' (Government of Telangana 2015, 11). Rajasthan's policy also mentions that the solar power producer shall contribute a sum of ₹25,000 per MW towards the Local Area Development Fund on a one-time basis (Government of Rajasthan 2019, 16). Among the remedial measures, there is an overwhelming emphasis on monetary compensation, while rehabilitation and resettlement are not explicitly mentioned. Monetary compensation can be inadequate because it does not account for appreciation in land value, the importance of land as a source of employment and its role in the socio-cultural dimension of people's lives (Maitra 2009; Yenneti and Day 2015). In simpler terms, one-time compensations cannot substitute for long-term losses of livelihood, and while compensatory processes involve some community consent and participation, they are far from just.

A few policies make bolder attempts to protect the rights of communities. Kerala's policy makes several provisions for the use of tribal lands, such as: 'The willingness of the land owner is mandatory'; 'The land ownership rights shall continue to fully vest with the original owner. The developer shall have only rights to setup and operate the project. The landowner will have the right to use land for agricultural purpose'; and 'Revenue (not profit) sharing based on the power generated, possibly in the range not below of 5% is envisaged' (Government of Kerala 2013, 8). The West Bengal policy is one of the few offering specific guidance on earmarking compensation for rehabilitation and resettlement purposes through the clause, 'Developer acquiring land must provide money (1% of project cost) to rehabilitate and resettle displaced people, for local development activities like building schools' (Government of West Bengal 2012, 32). The West Bengal and Jammu and Kashmir policies are notable for having a separate section on social and environmental issues (Department of Science and Technology 2013). Most policies, however, limit themselves to the technical and financial details of implementation.

The bigger picture

India's 100-GW grid-connected solar target consists of sub-targets for large- or medium-scale solar (60 GW) and distributed solar (40 GW). In practice, the vast majority of realized capacity is in the form of large-scale plants. By the end of 2019, India had 35.7 GW of solar capacity, of which only 4.4 GW was rooftop solar (Sanjay 2020). This suggests that India is swiftly moving towards a system configuration where utility-scale solar (and wind) will replace large thermal generators while retaining the existing institutional and political structure of the energy system. Decentralized energy systems, and their potentially emancipatory politics, are likely to get sidelined if these trends continue.

Most states resort to presenting large-scale solar parks and decentralized solar as different options, modes, models, or categories of projects. Some states present MW targets for decentralized capacity. However, on the whole, policy documents shy away from choosing between centralized and decentralized typologies. This 'all-of-the-above' approach reveals that the key priority for states is rapidly increasing deployment, irrespective of how it happens. Delhi stands out by framing its solar policy explicitly around rooftop solar, but this is perhaps only because of the limited space available for utility-scale solar in Delhi.

Further, decentralization alone is not sufficient to ensure community ownership as envisioned by energy democracy scholars. Within the rooftop segment, for example, the renewable energy service company (RESCO) model, where the developer retains ownership of the solar installation, constitutes 35 per cent of the rooftop capacity and is gaining steam (Bridge to India 2019; CII 2019). While some state policies, like Punjab, Jharkhand, Odisha, and West Bengal, mention increasing community participation in the electricity sector, none of them provides a mechanism to ensure increased *public ownership* of energy infrastructure (Government of Odisha 2013; Government of Punjab 2012). This question is partly engaged with in the Kerala policy, which states that 'a wider community ownership model with direct financial stake by the public shall be encouraged' for a niche segment – floating solar plants and public place installations (Government of Kerala 2013, 6).

Our survey of state RE plans, alongside the installed capacity numbers, suggests that India is in the process of reconfiguring its energy system – in terms of scale, ownership, and spatial spread – in line with the existing system. A push to ensure community ownership and control is almost completely missing from political discourse. While we do not wish to uncritically advance decentralized systems as the normative choice for India, we do intend to highlight that a monumental political process is underway right now without much public deliberation. The outcome of this process may lock in institutional effects that limit a just and democratic energy transition.

Conclusion

Our analysis focuses on solar power policies at the state level, given the salience of solar energy in India's current drive to realize an energy transition. We find that while energy justice concerns are not the core of state solar policies, there are innovative provisions in some of them that could create a more fair and participatory system if scaled widely. While this is a critical first step, research on questions of energy justice and democracy is nascent in India and several opportunities for further work exist. Future work in this space can develop in two directions.

First, from an empirical perspective, our analysis is limited to solar energy transition given the significance of this resource in India's current RE discourse. Similar distributive and procedural justice frames can be applied to other energy sources and forms of energy use (transportation, heating, cooking). Other sources and uses vary in their levels of complexity, organizational and institutional architecture, and resources required for their uptake. This could yield more nuanced insights on planning for just and democratic transitions. Second, more fundamental processes of democratic participation in the Global South need to be theoretically explored in the context of energy. Our framework largely refers to policy processes, but makes some fundamental assumptions about how and why people participate in democratic processes and the co-production of public services. The literature on coproduction is still nascent in the Global South and has the potential to offer insights into whether and under what conditions individuals and groups will be willing to own and manage complex public infrastructure.

The goal of our analysis has been to bring into focus the broader injustices and political visions for India's RE transition. This is by no means discounting the historic impetus of increasing energy access and sufficiency. Rather, we wish to reframe what radical success looks like in the Indian energy sector, both from a developmental and a climate mitigation point of view. Achieving multiple objectives (access, social justice, job creation, and affordable power) simultaneously is the only way to develop sustainably. This requires critically evaluating whether our energy politics, especially our RE politics, can truly achieve our stated developmental and social goals beyond decarbonization. Bringing in greater justice and democracy in the energy discourse serves as an entry point for this exercise.

Acknowledgements

We would like to thank the editor of this volume, Prakash Kashwan, for giving us the opportunity to work on this important area of research. We have benefitted immensely from his feedback. We also thank Akila Ranganathan and Rohit M. from Ashoka University for their exceptional research assistance throughout the writing process.

References

Aklin, M., P. Bayer, S. P. Harish, and J. Urpelainen. 2017. 'Does Basic Energy Access Generate Socioeconomic Benefits? A Field Experiment with Off-Grid Solar Power in India'. *Science Advances* 3 (5): e1602153. https://www.science.org/doi/pdf/10.1126/sciadv.1602153.

Alstone, P., D. Gershenson, and D. M. Kammen. 2015. 'Decentralized Energy Systems for Clean Electricity Access'. *Nature Climate Change* 5 (4): 305–314.

Amin, A. 2014. 'Lively Infrastructure'. *Theory, Culture and Society* 31 (7–8): 137–161.

Angel, J. 2016. *Towards Energy Democracy: Discussions and outcomes from an International Workshop* (Workshop Report). Amsterdam: Transnational Institute.

Banerjee, A., E. Prehoda, R. Sidortsov, and C. Schelly. 2017. 'Renewable, Ethical? Assessing the Energy Justice Potential of Renewable Electricity'. *AIMS Energy* 5 (5): 768–797.

Baruah, B. 2015. 'Creating Opportunities for Women in the Renewable Energy Sector: Findings from India'. *Feminist Economics* 21 (2): 53–76.

Baviskar, A. 1995. *In the Belly of the River: Tribal Conflicts over Development in the Narmada Valley*, 1st edn. Delhi: Oxford University Press.

Bhattacharyya, S. C. 2013. 'To Regulate or Not to Regulate Off-Grid Electricity Access in Developing Countries'. *Energy Policy* 63 (December): 494–503.

Bridge, G., B. Özkaynak, and E. Turhan. 2018. 'Energy Infrastructure and the Fate of the Nation: Introduction to Special Issue'. *Energy Research & Social Science* 41 (July): 1–11.

Bridge to India. 2019. *India Solar Rooftop Market*. Gurgaon: Bridge to India.

Burke, M. J. 2018. 'Shared Yet Contested: Energy Democracy Counter-Narratives'. *Frontiers in Communication* 3 (22). https://doi.org/10.3389/fcomm.2018.00022.

Burke, M. J. and J. C. Stephens. 2017. 'Energy Democracy: Goals and Policy Instruments for Sociotechnical Transitions'. *Energy Research & Social Science* 33: 35–48.

Candel, J. J. L. 2019. 'The Expediency of Policy Integration'. *Policy Studies* 42 (4): 1–16.

Candel, J. J. L. and R. Biesbroek. 2016. 'Toward a Processual Understanding of Policy Integration'. *Policy Sciences* 49 (3): 211–231.

Chaurey, A., P.R. Krithika, D. Palit, S. Rakesh, and B. K. Sovacool. 2012. 'New Partnerships and Business Models for Facilitating Energy Access'. *Energy Policy* 47 (June): 48–55.

Chunekar, A. and A. Sreenivas. 2019. 'Towards an Understanding of Residential Electricity Consumption in India'. *Building Research and Information* 47 (1): 75–90.

CII. 2019. *Powering your Rooftops* 2019. New Delhi: Confederation of Indian Industry (CII).

D'Souza, R. 2019. 'Should Clean Energy be Politics As Usual?' *Asian Affairs* 4 (2): 38–44.

D'Souza, S. 2019. '100% Electrification: Assessing Ground Reality'. Brookings, 2 April. https://www.brookings.edu/opinions/100-electrification-assessing-ground-reality/ (accessed 7 December 2021).

Debnath, R., G. M. F. Simoes, R. Bardhan, S. M. Leder, R. Lambert, and M. Sunikka-Blank. 2020. 'Energy Justice in Slum Rehabilitation Housing: An Empirical Exploration of Built Environment Effects on Socio-Cultural Energy Demand'. *Sustainability* 12 (7): 3027.

Department of Science and Technology. 2013. 'Solar Power Policy for Jammu & Kashmir'. https://jakeda.jk.gov.in/links/Solar Policy for J&K.pdf (accessed 21 February 2020).

Dominic Mathews, R., S. Barria, and A. Roy. 2016. 'Up from Development: A Framework for Energy Transition in India (No. 8)'. https://rosalux.nyc/wp-content/uploads/2021/03/tuedworkingpaper8.pdf (accessed 8 December 2020).

Dubash, N. K., S. S. Kale, and R. Bharvirkar. (eds.). 2018. *Mapping Power: The Political Economy of Electricity in India's States Delhi*: Oxford University Press.

Dubash, N. K., A. K. Swain, and P. Bhatia. 2019. 'The Disruptive Politics of Renewable Energy'. The India Forum, May.

Ganguly, R., R. Jain, K. R. Sharma, and S. Shekhar. 2020. 'Mini Grids and Enterprise Development: A Study of Aspirational Change and Business Outcomes among Rural Enterprise Owners in India'. *Energy for Sustainable Development* 56: 119–127.

Government of Assam. 2018. 'Assam Solar Energy Policy, 2017'. https://www.aegcl.co.in/06022018Gazette-Solar_Policy.pdf (accessed 5 February 2020).

Government of Bihar. 2017. 'Bihar Policy for Promotion of New and Renewable Energy Sources 2017'. http://energy.bih.nic.in/docs/renewable-energy-sources-2017.pdf (accessed 5 February 2020).

Government of Gujarat. 2015. 'Gujarat Solar Power Policy – 2015'. https://www.solartiger.in/SolarPolicies/GUJARAT__SOLAR POLICY.pdf (accessed 5 February 2020).

Government of Himachal Pradesh. 2016. 'H.P. Solar Power Policy – 2016'. http://www.cbip.org/Policies2019/PD_07_Dec_2018_Policies/Himachal Pradesh/1-Solar/2 Order _Solar_Policy. pdf (accessed 5 February 2020).

Government of Jharkhand. 2018. 'Jharkhand State Solar Rooftop Policy 2018'. at https://www.jreda.com/upload_files/kvit131.pdf (accessed 6 February 2020).

Government of Karnataka. 2014. 'Karnataka Solar Policy 2014–2021'. http://kredlinfo.in/solargrid/Solar Policy 2014-2021.pdf (accessed 5 February 2020).

Government of Kerala. 2013. 'Kerala Solar Energy Policy 2013'. https://kerala.gov.in/documents/10180/46696/Kerala Solar Energy Policy 2013 (accessed 6 February 2020).

Government of NCT of Delhi. 2016. 'Delhi Solar Policy 2016'. http://ipgcl-ppcl.gov.in/documents/renewable/2016_08_03_6_Delhi_Solar_Policy.pdf (accessed 5 February 2020).

Government of Odisha. 2013. 'Odisha Solar Policy-2013'. http://www.cbip.org/Policies2019/ PD_07_Dec_2018_Policies/Orissa/1-Solar/2 Order Odisha-Solar-Power-Policy.pdf (accessed 5 February 2020).

Government of Punjab. 2012. 'New and Renewable Sources of Energy (NRSE) Policy – 2012'. https://www.peda.gov.in/media/pdf/nrse pol 2012.pdf (accessed 20 February 2020).

Government of Rajasthan. 2019. 'Solar Energy Policy, 2019'. https://jalore.rajasthan.gov.in/ content/dam/doitassets/jalore/pdffiles/Rajasthan Solar Energy Policy2019.pdf (accessed 2 June 2020).

Government of Sikkim. 2019. 'Grid Connected Rooftop Solar Photovoltaic System Policy for Sikkim – 2019'. https://power.sikkim.gov.in/wp-content/uploads/2019/10/SIKKIM-GAZETTE.docx (accessed 20 February 2020).

Government of Telangana. 2015. 'The Telangana Solar Power Policy 2015'. http://industries. telangana.gov.in/Library/TS Solar Policy.pdf (accessed 5 February 2020).

Government of West Bengal. 2012. 'Policy on Co-generation and Generation of Electricity from Renewable Sources of Energy'. http://www.cbip.org/Policies2019/PD_07_Dec_2018_ Policies/West Bengal/1 Summary West Bengal Renewable Policy - 2012.pdf (accessed 6 February 2020).

Government of Telangana. 2015. 'The Telangana Solar Power Policy 2015'. http://industries. telangana.gov.in/Library/TS Solar Policy.pdf (accessed 5 February 2020).

GPI Secretariat. 2016. 'GHG Platform India'. http://www.ghgplatform-india.org/ (accessed 8 December 2021).

Guruswamy, L. 2011. 'Energy Poverty'. *Annual Review of Environment and Resources* 36 (1): 139–161.

Heffron, R. J. and D. McCauley. 2017. 'The Concept of Energy Justice across the Disciplines'. *Energy Policy* 105: 658–667.

Hughes, L. and P. Y. Lipscy. 2013. 'The Politics of Energy'. *Annual Review of Political Science* 16 (1): 449–469.

Jenkins, K., D. McCauley, R. Heffron, H. Stephan, and R. Rehner. 2016. 'Energy Justice: A Conceptual Review'. *Energy Research & Social Science* 11: 174–182.

Kohli, K. 2013. 'Two Coal Blocks and a Political Story'. *Economic and Political Weekly* 48 (41): 12–14.

Lacey-Barnacle, M., R. Robison, and C. Foulds. 2020. 'Energy Justice in the Developing World: A Review of Theoretical Frameworks, Key Research Themes and Policy Implications'. *Energy for Sustainable Development* 55: 122–138.

Larkin, B. 2013. 'The Politics and Poetics of Infrastructure'. *Annual Review of Anthropology* 42 (1): 327–343.

Maitra, S. 2009. 'Development Induced Displacement: Issues of Compensation and Resettlement –Experiences from the Narmada Valley and Sardar Sarovar Project'. *Japanese Journal of Political Science* 10 (2): 191–211.

Malakar, Y., M. J. Herington, and V. Sharma. 2019. 'The Temporalities of Energy Justice: Examining India's Energy Policy Paradox Using Non-Western Philosophy'. *Energy Research and Social Science* 49: 16–25.

McCauley, D., R. J. Heffron, H. Stephan, and K. Jenkins. 2013. 'Advancing Energy Justice: The Triumvirate of Tenets'. *International Energy Law Review* 32 (3): 107–110.

Moallemi, E. A., L. Aye, J. M. Webb, F. J. de Haan, and B. A. George. 2017. 'India's On-Grid Solar Power Development: Historical Transitions, Present Status and Future Driving Forces'. *Renewable and Sustainable Energy Reviews* 69: 239–247.

Monyei, C. G., K. E. H. Jenkins, C. G. Monyei, O. C. Aholu, K. O. Akpeji, O. Oladeji, and S. Viriri. 'Response to Todd, De Groot, Mose, McCauley and Heffron's Critique of "Examining Energy Sufficiency and Energy Mobility in the Global South through the Energy Justice Framework"'. *Energy Policy* 133: 110917.

Monyei, C. G., K. E. H. Jenkins, V. Serestina, and A. O. Adewumi 2018. 'Examining Energy Sufficiency and Energy Mobility in the Global South through the Energy Justice Framework'. *Energy Policy* 119: 68–76.

Munro, P., G. van der Horst, and S. Healy. 2017. 'Energy Justice for All? Rethinking Sustainable Development Goal 7 through Struggles over Traditional Energy Practices in Sierra Leone'. *Energy Policy* 105: 635–641.

Palit, D., B. K. Sovacool, C. Cooper, D. Zoppo, J. Eidsness, M. Crafton, et al. 2013. 'The Trials and Tribulations of the Village Energy Security Programme (VESP) in India'. *Energy Policy* 57: 407–417.

PMO India. 2019. 'PM Modi addresses Climate Action Summit'. Press Information Bureau, September. https://pib.gov.in/Pressreleaseshare.aspx?PRID=1585979 (accessed 8 December 2021).

Sanjay, P. 2020. 'Mercom Reveals India's Solar Market Leaders in CY 2019'. *Mercom India*, April.

Sharma, K. R., D. Palit, P. Mohanty, and M. Gujar. 2014. 'Approach for Designing Solar Photovoltaic-based Mini-Grid Projects: A Case Study from India'. In *Mini-Grids for Rural Electrification of Developing Countries*, edited by S. C. Bhattacharyya and D. Palit. Cham: Springer, 167–201.

Sharma, K. R., D. Palit, and P. K. Ramakrishnan. 2016. 'Economics and Management of Off-Grid Solar PV System'. *In Solar Photovoltaic System Applications*, edited by P. Mohanty, T. Muneer, and M. Kolhe. Cham: Springer International Publishing, 137–164.

Sharma, K. R., and D. Palit. 2020. 'A Coevolutionary Perspective on Decentralised Electrification: A Solar Mini-Grid Project in India'. *International Journal of Sustainable Society* 12 (3): 202–219.

Shidore, S., and J. W. Busby. 2019. 'What Explains India's Embrace of Solar? State-led Energy Transition in a Developmental Polity'. *Energy Policy* 129: 1179–1189.

Sovacool, B. K. and M. H. Dworkin. 2015. 'Energy Justice: Conceptual Insights and Practical Applications'. *Applied Energy* 142: 435–444.

Stephens, J. C. 2019. 'Energy Democracy: Redistributing Power to the People through Renewable Transformation'. *Environment: Science and Policy for Sustainable Development* 61 (2): 4–13.

Sweeney, S. 2012. 'Resist, Reclaim, Restructure: Unions and the Struggle for Energy Democracy', Discussion document, October. Trade Unions for Energy Democracy, New York.

Szulecki, K. 2018. 'Conceptualizing Energy Democracy'. *Environmental Politics* 27 (1): 21–41.

Todd, I., J. De Groot, T. Mose, D. McCauley, and R. J. Heffron. 2019. 'Response to "Monyei, Jenkins, Serestina and Adewumi Examining Energy Sufficiency and Energy Mobility in the Global South through the Energy Justice Framework"'. *Energy Policy* 132: 44–46.

University of Wisconsin Center for Cooperatives. 2020. 'Rural Electric Cooperatives'. https://mce. uwcc.wisc.edu/utilities-overview/rural-electric-cooperatives/ (accessed 16 July 2020).

Willcox, M., A. Pueyo, L. Waters, R. Hanna, H. Wanjiru, D. Palit, and K. R. Sharma. 2015. 'Utilising Electricity Access for Poverty Reduction'. Practical Action Consulting (PAC), Institution of Development Studies (IDS), and The Energy and Research Institute (TERI), Rugby, UK.

Winkler, H., A. F. Simões, E. L. La Rovere, M. Alam, A. Rahman, and S. Mwakasonda. 2011. 'Access and Affordability of Electricity in Developing Countries. *World Development* 39 (6): 1037–1050.

Yanow, D. 2007. 'Interpretation in Policy Analysis: On Methods and Practice'. *Critical Policy Studies* 1 (1): 110–122.

Yenneti, K. and R. Day. 2015. 'Procedural (In)justice in the Implementation of Solar Energy: The Case of Charanaka Solar Park, Gujarat, India'. *Energy Policy* 86: 664–673.

CHAPTER 4

Extractive Regimes in the Coal Heartlands of India

Difficult Questions for a Just Energy Transition

Vasudha Chhotray

Introduction

It is now accepted that the future of coal will be decided in the developing world. Even as Western countries transition away from coal, increased production and consumption of coal in India and China have meant that the share of coal in global energy production has remained constant for the past 40 years, despite attempts at decarbonization (Edwards 2019). Nevertheless, the West continues to produce high per capita emissions compared to developing nations (Lazarus and van Asselt 2018). In response, India has asserted its rights to equitable energy access in the international arena (Jaitly 2021). At the same time, questions of intra-country equity complicate India's position, with many arguing that India must pursue low-carbon pathways to protect its poor and vulnerable groups (Bidwai 2012).

After Independence, coal became an enduring symbol of national development in India (Lahiri-Dutt 2014). The coal industry has deep political roots, engaging powerful stakeholders at different levels (Bhattacharjee 2017). In recent years, coal investments have lost their appeal due to unrest over their environmental impacts as well as a dynamic downward trend in the demand for thermal power (Rajshekhar 2021). Even so, production targets for the state-owned Coal India Limited (CIL) – responsible for over 80 per cent of India's coal production – were increased to 1 billion metric tonnes by 2024. The central government is actively looking to sell more coal blocks to raise money, despite the lukewarm response to recent coal block auctions. Coal imports have simultaneously increased, engendering a new coastal coal geography controlled by private actors (Oskarsson et al. 2021). That renewables cannot substitute for coal, despite policy support from the state, is accepted. Analysts expect coal-fired generation to continue to grow to meet electricity demand growth even if 350 gigawatt (GW) of renewable energy (RE) capacity is installed by 2030 (Tongia and Gross 2019). New energy forms, including renewables, are, historically speaking, energy 'additions' rather than 'transitions' (Oskarsson et al. 2021). Importantly, this perception is not typical of India alone, as the global energy system remains locked into high coal energy use in the midst of an RE boom (Oskarsson et al. 2021).

Energy transitions worldwide are characterized by incremental change on the part of states and private sectors rather than radical transformation (Newell 2019). In India, too, recent moves to decarbonize through state patronage of RE, that is, solar and wind projects, have reinforced, not disrupted, the logics of 'fossil developmentalism' (Chatterjee 2020, 3). Large-scale solar and wind projects cause displacement and environmental damage and provoke resistance (Stock and Birkenholtz 2019). They generate little local employment and instead make inhabitants' livelihoods more precarious through the dispossession of common lands (Yenneti, Day, and Golubchikov 2016).

A just transition to a low-carbon future would require addressing political and ethical questions and paying attention to the interlinked issues of equity and justice. In India, nearly 400 million people still lack access to electricity, and average urban emissions per capita are 2.5 times higher than in rural areas (Chakravarty and Ramana 2012). Many of these energy-deprived people live in the coal heartlands of Jharkhand and Chhattisgarh, two of the poorest states in India.

A Gond Adivasi activist, in the Hasdeo Arand reserve, Chhattisgarh, said,

> They use our coal to generate electricity and the shame is that we only recently got it two and a half years ago. They say that they cannot give us a railway or telephone line because the forest is too dense, yet there is now a coal train that runs through the forest, all day, every day. (Sra 2020)

These 'frontier' communities powerfully illustrate that the burden of climate debt is borne unequally in the Global South, where many poor and indigenous people are exploited by the carbon economy. There are complex interdependencies between local livelihoods and the coal industry (Lahiri-Dutt 2014; Noy 2020), especially within sectors where unionized workers are a minority (Chandra 2018). These conditions make it hard to mimic Western experiments with just transitions – for instance, the Polish experience where the coal union was actively involved in effecting the transition (Zinecker et al. 2014).

In recent years, scholars have argued that just transitions must account for the futures of fossil-fuel workers (Newell and Mulvaney 2013). There has been scant research on specific challenges associated with the extractive industry in the Indian coal heartlands, but it is emerging more clearly as an area of concern (Pai, Harrison, and Zerriffi 2020; Bhushan, Banerjee, and Agarawal 2020). To secure coal miners' livelihoods, India needs to significantly scale up its solar capacity in coal mining areas so that they can find employment in the new sector (Pai et al. 2020). This relates to macro-level discussions on the extent to which RE projects could make coal less relevant in the future (Rajshekhar 2020). However, energy transitions are not apolitical matters restricted to 'technical' choices around fuels or energy technologies (Bridge and Gailing 2020; Newell 2019). All energy interventions potentially reconfigure a broad field of social and political power within historically unequal spaces. There remains a deficit of critical research at the sub-national level, which is a massive omission in the case of India, as the bulk of focus on coal mining and potential RE projects unfolds in a few states around the country.

In this chapter, I offer an analytical framework to examine sub-national 'extractive' regimes that shape the discursive, institutional, and political context within which extraction is being organized. Drawing on the cases of Jharkhand and Chhattisgarh, two of India's principal coal-producing states, I argue that the concept of extractive regimes can identify the drivers of continuing injustices in coal mining today and illuminate the spaces, actors, and networks that are currently agitating for justice therein. These must be brought on board for bottom–up political engagement to steer a just transition for communities at the coal frontiers. This analysis will also shed light on potential enduring injustices that will be replicated in RE projects that are initiated in these spaces, as well as possible opportunities for change. The chapter draws upon critical qualitative research conducted from 2014 to 2017, including more than 100 interviews with key informants and relevant secondary sources, such as academic articles, policy documents, RTI information, and news reports.

Conceptual foundations: extractive regimes and the politics of a just transition

Charting a just transition for the Indian coal heartlands requires systematic engagement with the extractive regime and examining how a move away from coal might unfold. The notion of a 'regime' of power is useful to draw attention to both formal and informal structures of power, the vast configuration of actors and institutions, and, more generally, societal norms (Kashwan 2017). An 'extractive' regime of power specifically theorizes the discursive, institutional, and political apparatuses around extraction (Adhikari and Chhotray 2020).

In India, mineral resource governance is centralized, resting upon key national laws like the Coal Bearing Areas Act 1957, the Mines and Minerals (Development and Regulation) Act 1957 (MMDRA), and the Land Acquisition, Rehabilitation and Resettlement Act 2013 (LARRA), and the vital roles of the central ministries of coal and environment and forests. However, following economic liberalization and deregulation in 1991, states began competing among each other for economic investment. Coal, nationalized in 1973, was gradually opened to private sector involvement from 1993. Besides actively soliciting private investment in mining, states put in place effective institutional mechanisms for resettlement and compensation and to deal with local resistance, using coercion if necessary (D'Costa and Chakraborty 2017; Sud 2019). 'Broker states' that balance this duality of purpose (land acquisition via palliation but also crackdown) have memorably been theorized as 'regimes of dispossession' (Levien 2018, 4).

With RE development, a key focus of the policy push is encouraging private investment through tax benefits and capital subsidies (Lakhanpal 2019). The Central Electricity Act 2003 devolved RE policymaking to sub-national actors, and states have since been courting private investors by facilitating land acquisition, evacuating power, and building access roads. The states promoting RE projects aggressively are also those with highly liberalized land laws: Gujarat, Rajasthan, Tamil Nadu, and Andhra Pradesh. These are not in the coal heartlands.[1] India also mirrors the International Labour Organization's global prediction that 'green jobs' in the energy sector will be unequally distributed (Zinecker et al. 2014). However, the lack of new RE projects is not the only problem associated with achieving a just transition in the coal heartlands. Emerging research from states where wind and solar projects are situated unequivocally points to new forms of dispossession for poor people, driven by state-promoted large-scale land acquisition (Stock and Birkenholtz 2019).

The switch from fossil fuels to renewable projects needs to be viewed as a continuum that unfolds within an existing extractive regime. If energy transitions are the 'production of novel combinations of energy systems and social relations across space' (Bridge and Gailing 2020, 1038), then it is not just the kind of technology that

---

[1]   Strong Adivasi constituents exert a counter pressure on excessively liberalizing land laws in Jharkhand and Chhattisgarh. More on their extractive regimes in the section title 'Extractive regimes in the coal heartlands of India: Jharkhand and Chhattisgarh'.

is deployed for energy production, but the type of extractive regime that oversees the entire operation that matters for realizing a just transition. To fully appreciate the political economy required to achieve a just energy transition in India, we need to engage with the historical 'extractive imperative', which provides the ideological basis for states to promote extractive projects (Arsel, Hogenboom, and Pellegrini 2016). In this chapter, I discuss three critical dimensions that make up an extractive regime: the public, political legitimation of extraction, institutional effectiveness, and the management of resistance.

Background: the nested injustices of coal in India

Coal mining led to the creation of a distinctive class of workers, comprising primarily rural and Adivasi migrants, in the early colonial collieries, situated in Dhanbad (Bihar) and Raniganj (West Bengal). At Independence, the Constitution declared that subsoil minerals were state-owned[2] and the government enacted strong central laws that minimized the rights of local communities (Lahiri-Dutt 2016, 2014). After nationalization, coal mining stretched westwards, into forested territories mainly inhabited by Adivasis. The expansion of mining was accompanied by the uneven development of urban industrial tracts, destruction of local flora, and undermining of the agricultural economy (Oskarsson, Lahiri-Dutt, and Wennström 2019). Progressive laws in favour of Adivasis in coal-rich states like Jharkhand, Chhattisgarh, and Odisha have been legislated over the decades, though they are frequently sabotaged by vested interests (Sundar 2007).

The intensity of environmental devastation has worsened with the World Bank-backed shift from traditional underground mining to open-cast mining (Oskarsson, Lahiri-Dutt, and Wennström 2019). A particularly egregious effect of this shift was felt in Jharia, Dhanbad, where the abandoned underground mines were never filled with sand, thus allowing oxygen to enter through the seams to the burning coal below. Fires may occur in coal layers that are exposed to the surface of the earth, and Jharia has experienced coal fires since 1912, but clearly, the shift has only made things worse (Pai 2018). While CIL has not seriously pursued environmental protection and reclamation, state pollution boards have failed to enforce regulations meant to control fly ash disposal, stack emissions, and effluent wastewater treatment (Chandra 2018). Serious detrimental impacts of coal mining, transportation, waste, and combustion include air pollution and long-term damage to the ecosystem (CSE 2008).

---

[2] The exception is Meghalaya, where, by virtue of its special Sixth Schedule status, coal is owned by the indigenous communities who own the land.

The expansion of CIL led to an employment boom and the subsequent bolstering of trade unions throughout the coal belt. By 1965, almost 255,000 workers (about 60 per cent of the total workforce) were enrolled in unions (Chandra 2018). Powerful unions lobbied for greater spending from the Coal Mine Welfare Fund, a trend that continued after nationalization. In response to labour mobilization, many other benefits were rolled out, from healthcare to housing. Through the 'social multiplier effect', while CIL's formal workforce may have been around 650,000 people at its peak, the actual number of beneficiaries was much larger, as each CIL employee could add five members to their medical card. In some ways, CIL substituted for states in carrying out developmental functions in historically poor and disadvantaged parts of India (Chandra 2018).[3] Interestingly, the coal sector has a higher level of value-added tax and wages compared to other sectors (Spencer et al. 2018). However, sub-state data are scarce, not harmonized and, besides, they do not capture the income levels of informal coal workers.

The distribution of CIL benefits was controlled by a few union leaders who were backed by ruling party politicians. Contractors and sardars controlled labour recruitment in a perpetual 'shadow economy', employing informal sector workers at appallingly low wages. Moreover, safety protocols and compensation mechanisms are not followed even among formal CIL workers, and most contract workers' deaths are not even reported by labour contractors (Pai 2018). A notorious coal mafia has evolved that has a stranglehold over trade unions and mine labour (Singh and Harriss-White 2019). Practices of engaging informal labour who toil in extremely poor working conditions abound in private coal companies too (Lahiri-Dutt 2016). Thus, the power of many coal unions is declining.

A large artisanal mining sector also operates in and around these coal mines. Coal-cycle–*wallahs* scavenged coal in heavy sacks on bicycles to nearby market towns for sale (Pai 2018). However precarious and difficult, this is often considered preferable to other available kinds of informal labour. Such small-scale mining engages nearly 400,000 people in India and is regarded as illegal (Lahiri-Dutt 2016). Even though such perilous work is conducted illegally and on the margins, some people (including Adivasis) have been able to benefit from the informal coal economy and, indeed, there is a multi-level political nexus around the lucrative illegal transportation and sale of coal (Singh and Harriss-White 2019). The highly unequal workforce of the coal economy is experiencing ever new inequalities along the lines of class, caste, tribe, and gender (Noy 2020). Further, there has been little

---

[3] Private companies offer such functions via CSR (corporate social responsibility); more later.

concerted attempt by the state to invest the proceeds of mining back into these communities. MMDRA 2015 provides for district mineral foundations to channel a fraction of royalty payments and auction proceeds to local communities, but these remain marred by ambiguity and underutilization (Chakraborty, Garg, and Singh 2016; Banerjee 2020).

Amidst all this, coal extraction continues, and state-driven land acquisition perpetuates enduring injustice, which is met with changing forms of resistance (Sathe 2016). Levien (2013) suggests that Adivasis in remote mineral-rich forested areas are far less willing to accept compensation than their urban and peri-urban counterparts in big metropolitan centres. However, this does not present the full picture, as other research has shown that compensation is perceived as attractive and is sought after; it even triggers new patterns of differentiation between those whose lands have been directly dispossessed, rendering them eligible for compensation, and those affected by other factors, like loss of access to commons resources, but do not qualify (Kale 2020; Noy 2020).

Extractive regimes in the coal heartlands of India: Jharkhand and Chhattisgarh

In 2000, two of the country's richest coal-producing areas, Bihar and Madhya Pradesh, were bifurcated, yielding two new mineral-rich states, Jharkhand and Chhattisgarh. Incoming political elites had a valuable opportunity to govern these new states in alignment with their ideologies and broader interests (Adhikari and Chhotray 2020). They vigorously championed extractivist ideas of development and went on to promote mining, which contributed approximately 10 per cent towards each state's gross domestic product (GDP) in 2000–2014. For context, mining accounted for 1.2 per cent of the national GDP in both 2000 and 2010 (Chakraborty, Garg, and Singh 2016).

Jharkhand and Chhattisgarh represent old and new sites for coal mining, respectively. There are differences in the 'coal cultures' of particular CIL subsidiaries; for instance, political bargaining and negotiating are more established in older areas like Jharkhand (Chandra 2018). CIL is deeply intertwined with regional and local state apparatuses, and state governments are responsible for acquiring land for CIL.

In the rest of this section, I discuss the main dimensions of the extractive regimes of Jharkhand and Chhattisgarh: their political history and organization, which enable public, political commitment to extraction, and the institutional apparatus of each state, which facilitates extraction (including land acquisition) and effectively manages resistance. Both states have specific laws that relate to the governance of their extensive Scheduled Areas and the rights of Scheduled Tribes, but these are outside the scope of discussion here (Wahi and Bhatia 2018).

Political history and organization

Jharkhand was formed following more than half a century of political mobilization by Adivasi social movements, and the nurturing of a clear Adivasi political identity, with the Jharkhand Mukti Morcha (JMM) emerging as the leading political party. Although the expansion of the Bharatiya Janata Party's (BJP) influence on regional politics substantially de-linked statehood from Adivasi identity, it did not prevent the idea of Jharkhand as a homeland for tribal people from enduring both politically and in popular memory (Tillin 2013). This positioning, following from the legacy of statehood, has made it difficult for Jharkhand's political leadership to strike an appropriate public, political discourse about extractive development since 2000.

Even after attaining statehood, Jharkhandi identity has remained closely tied to the 'premise of resistance', articulated through ideological binaries like the tribal versus the non-tribal exploitative 'outsider' (Hebbar 2003). All political parties in the state, whether Adivasi or not, have had to engage publicly with the question of whether the new state serves the interests of Adivasis. However, Adivasis are not politically homogenous and many supported the BJP in the elections of 2014, following the expansion of the Rashtriya Swayamsevak Sangh (RSS)–led grassroots mobilization, but as part of a larger clientelist relationship (Kumar 2018).[4] Yet, the BJP's growing presence in Jharkhandi politics has not taken away from the Adivasi social base of Jharkhand's many parties, which resort to frenzy and resistance, especially concerning 'resource grab' issues (Rajalakshmi 2016). The irony is that the centrality of Adivasi issues in Jharkhandi politics has had little impact on the unbroken pursuit of extractive activity since the early 2000s.

As opposition leader, Hemant Soren of the JMM said, in 2015, that the party was not 'against industry', but that people should not be made to feel like they did not own the land and its resources (Adhikari and Chhotray 2020, 856). There has been relatively little 'domestic' participation in private mining in Jharkhand, compared to Chhattisgarh, with most entrants like Jindal and Rungta coming from outside the state. This has provided political ammunition to Jharkhandi Adivasi parties to rally against continued exploitation. Since coming to power in the last assembly elections in 2019, Soren has pledged to constitute a 'displacement commission to return

---

[4] Clientelism refers to the practice of distribution of benefits in exchange for electoral support.

unused land in possession of private and government agencies to their *raiyyats* (land right holders) as per the 2013 LARRA' (TNN 2020). Soren has also been openly critical of the Modi government's recent decision to auction 41 new coal blocks – many of which lie in Jharkhand – and for disregarding the state's concerns regarding Adivasi welfare and environmental costs (Barik 2020). However, the Supreme Court has accused Jharkhand's leadership of 'doublespeak' stating that they only sought a moratorium of six–nine months on the auctions so that they could secure higher bids following improvements in the global investment climate (Mahapatra 2020). Successive political parties have covertly facilitated land acquisition for private companies. State-sponsored violence, manipulation, and crushing of dissent through force are on the rise (Adhikari and Chhotray 2020; Choudhury 2018).

A comparable long-standing Adivasi-led statehood movement is missing in Chhattisgarh, where the ascendant Other Backward Classes (OBC) politics of the 1990s and political bargaining at the Centre delivered the new state (Berthet 2011). Despite the mobilization around Adivasis' right to *jal–jangle–zameen* (water–forest–land) since colonial times in Bastar (Sundar 1997), there is still only a tenuous link between Adivasi identity and mainstream political discourse in contemporary Chhattisgarh. This is further compounded by the absence of Adivasi political parties, which leaves the upper-caste-dominated Congress and BJP to compete for power. Both parties try to keep the OBC elite in check while also trying to appease various OBC lobbies (Adhikari and Chhotray 2020). Moreover, grassroots mobilization and service delivery by RSS-affiliated organizations, like the Vanavasi Kalyan Ashram, have contributed to the BJP gaining Adivasi support in Chhattisgarh (Thachil 2014).

Following the creation of a non-Adivasi 'Chhattisgarhi middle class,' comprising upper castes like Rajputs and Banias, an increasing share of private capital in mining comes from within the state (Adhikari and Chhotray 2020). Significant representation in successive governments, moreover, has ensured that upper-caste people, many of whom are traders and businessmen, now also command industrial capital – for example, through the ownership of power plants (Das Gupta 2019). This alliance between ruling political elites and rich upper castes has resulted in the public legitimation of extraction and state condonement of excesses by mining companies (Das Gupta 2019). A key element of this political discourse around extraction is the prominent silencing of any critical opposition of the ruling regime.

The comparatively left-of-centre Congress party has struggled to distinguish itself from the BJP. A senior Congress leader explained that it was 'unviable' for the party to oppose land acquisition for mining for fear of being labelled 'anti-development' (Adhikari and Chhotray 2020). In early 2019, the newly elected Congress government nevertheless took the bold step of returning land acquired by the previous BJP government for a Tata Steel plant to Adivasis in Bastar (CNBC-TV18 2019). The move was symbolically powerful, but the Congress government in the state reportedly continues to work with the pro-BJP Adani Group in several blocks of the coal-rich Korba district. This is hypocritical given the Congress' criticisms of previous BJP-led governments for facilitating backdoor corporate entry into environmentally sensitive regions. The state government has also permitted Adani enterprises to set up mining operations in the conflict-ridden, Adivasi-dominated Dantewada district in the south of the state (Sharma 2019). These developments strongly indicate that the Congress in Chhattisgarh is not likely to depart substantially from historically anti-tribal models of extractive development.

Institutional effectiveness

Institutional effectiveness is defined here from the perspective of the extractive industry, representing the institutional might of the state in facilitating approvals and acquisitions as well as orchestrating effective compensation policies so that mining projects can take off. It is not a measure of inclusivity though it offers a clearer definition of the rights of those impacted in various ways by mining (Adhikari and Chhotray 2020).

Jharkhand's administrators were candid about the apathetic institutional functioning in their state. Many officials confirmed that during the period of political instability between 2000 and 2013, basic administrative and monitoring procedures were neglected. However, there were relative improvements after the state got its first full-term government in 2014–2019 (see Adhikari and Chhotray 2020 for more details). Jharkhand's early years were adversely affected by inexperienced political leadership and a highly conflict-ridden bureaucracy that carried forward internal squabbles that had prevailed in its parent state, Bihar. The quick turnover of politicians resulted in exceptionally high transfer rates for key bureaucrats, primarily to ensure that pliant bureaucrats oversaw high rent-yielding sectors like industry and mines (Adhikari and Chhotray 2020). This was inimical to good administration.

Jharkhand inherited a lackadaisical industrial policy from Bihar, which, in the final years before its bifurcation, suffered an astonishing period of industrial decline under Laloo Prasad Yadav (Kale and Mazaheri 2014). The new state, with its high degree of political fragmentation and administrative paralysis, could not develop any new industrial policies until 2012. Indeed, there were no clear rehabilitation and resettlement policies in Jharkhand until 2015. The state's bureaucracy was grossly understaffed and notorious for its inertia. As multiple field studies continue to demonstrate, this institutional lethargy held no advantage for Adivasis facing land dispossession; on the contrary, land acquisition and Adivasi dispossession have continued, marked by irregularities, deception, and blatant misuse of pro-tribal land laws (Sundar 2007).

Unlike Jharkhand, which inherited a debilitated industrial sector from undivided Bihar, Chhattisgarh benefitted from the legacy of a stronger public sector undertakings (PSU)–led industrial policy in Madhya Pradesh (Adhikari and Chhotray 2020). Despite many challenges, its administrators succeeded in setting up a cohesive and competent team under the leadership of the first chief minister (CM), Ajit Jogi, a former Indian Administrative Service (IAS) official. They speedily divided their state administrative cadres within two years – a process that took double that time in Jharkhand – and were much more meticulous about staffing procedures, which led to fewer instances of understaffing (Adhikari and Chhotray 2020). The state's institutional capacity complemented its political enthusiasm for extraction, yielding a cohesive and coherent extractive regime in the state.

Under the long BJP rule from 2003 to 2019, the Chhattisgarh government took several concerted steps to effectively institutionalize the facilitation of extraction, for example, holding high-level meetings of the State Investment Promotion Board (SIPB) and creating a digitized land records database (Adhikari and Chhotray 2020). Between 2000 and 2015, Chhattisgarh signed 19 leases for coal, whereas Jharkhand signed 10. Moreover, Chhattisgarh's proactive state government regularly lobbied New Delhi to expedite central clearances, earning much appreciation from private actors who were dissatisfied with the government in Jharkhand (Adhikari and Chhotray 2020). Still, Chhattisgarh is not immune to the difficulties associated with land acquisition and has faced problems getting new projects started (Rajshekhar 2012).

State management of resistance

There are three broad areas of difference between the extractive regimes of Jharkhand and Chhattisgarh where state management of resistance is concerned. First, violent left-wing extremists or Naxals have resisted extraction in both states since the 1990s. In Jharkhand, such activity is geographically dispersed throughout the state; however, it is concentrated in a few southern districts like Bastar and Dantewada in Chhattisgarh. Both states have cracked down on extremist activity, though Chhattisgarh's notorious Salwa Judum is a testament to the state's superior institutional capacity to respond harshly. The Congress and the BJP both support this highly controversial vigilante army in Chhattisgarh. Jharkhand's own counter-response has been weaker and more diffused in contrast, although, in 2011, Operation Green Hunt led to a worrying increase in human rights violations and arrests of

Maoist 'sympathizers' (Adhikari and Chhotray 2020). Many Adivasi politicians and Naxal activists are deeply complicit in the coal trade and receive patronage from political parties, which complicates things further (Kumar 2018; Shah 2006).

Second, Jharkhand has a large, active network of social movement organizations that have been campaigning for Adivasi rights to land and forest resources. The Jharkhand Mines Area Coordination Committee (JMACC), a prominent and well-connected alliance of mining-affected communities, campaigns against irregularities and corruption in extractive development processes, organizes citizen tribunals, and demands compensation. The Chhattisgarh Mukti Morcha (CMM), a trade union movement from the 1970s, and the Ekta Parishad, a pan-Indian grassroots movement, have anchored civil society mobilization in the state, though both have waned over time. The Chhattisgarh Bachao Andolan (CBA) agitates against coal mining, forges solidarity among smaller organizations, and is well connected beyond the state. However, public protests are extremely difficult to organize in Chhattisgarh given the state's systematic silencing of such events (Das Gupta 2019) and broader civil society mobilization is curtailed through state repression. Though arrests of peaceful civil activists take place in both states, data from the South Asia Terrorism Portal affirms that this number is higher in Chhattisgarh (Adhikari and Chhotray 2020). The rise of corporate-owned media further enables the stifling of dissent, especially in Chhattisgarh (Adhikari and Chhotray 2020)

Third, Jharkhand's many anti-dispossession movements align themselves clearly with Adivasi political parties (Kumar 2018). Prominent activists associated with JMACC and the Jharkhand Organization for Human Rights (JOHAR) have collaborated with such parties in resisting mining. A strength of Jharkhand's plural political landscape is that it is not dominated by national parties with a centralized political culture like Chhattisgarh. This has allowed local, elected political representatives to support, join, and even lead acts of resistance. Given that the national extractive imperative shows no signs of slowing down and is acquiring ever harsher overtones, these local acts are becoming increasingly significant.

The micro-politics of historical injustices: links with extractive regimes

Coal companies are formidable political actors, driving many unfair practices within the coal heartlands (Chandra 2018; Lahiri-Dutt 2014). This section presents some vignettes from fieldwork carried out in 2015–2016 in the coal-rich districts of Hazaribagh in Jharkhand and Korba and Raigarh in Chhattisgarh. The focus is on understanding the dynamics around resistance while identifying links with sub-national extractive regimes.

The case of the CIL subsidiary CCL (Central Coalfields Limited) in Hazaribagh is important because it is part of an established coal history. Hazaribagh is situated in the North Karanpura Coalfield, where large-scale open-cast mining has been ongoing since the 1980s and underground mining even earlier. Planning records barely mention the profound social contestation of changing land use to accommodate extraction (Oskarsson, Lahiri-Dutt, and Wennström 2019). Regardless of official documentation, mining plans were calibrated to manage or neutralize resistance. There are no ongoing disputes against CCL in Hazaribagh, but official silence regarding past injustices is concerning.

National Thermal Power Corporation (NTPC), another leading PSU, has been persuading locals in Hazaribagh to give up their lands since launching operations in 2006. It claimed to use 'village mobilizers' for the purpose but faced staunch resistance. A local ex-Congress politician took up the cause on behalf of the agitating locals and accused senior district administrators and the police of being complicit in supporting malpractice. Leading anti-mining activists negotiated an acceptable rate for land sales via the office of the district collector, although the fact that the land had not been valued properly in the first place made the process harder. Conflict erupted, resulting in tragic police firing. Other planned acquisitions, also by CCL, in other parts of Jharkhand reportedly provoked massive protests as well (Yadav 2013).

Jharkhand's extractive regime must negotiate messy and robust forces of popular resistance. The recent, qualified move by CM Soren in support of the Centre's decision to facilitate commercial coal mining is itself indicative of ideological tightrope walking. Soren asked for a brief moratorium to create a policy balance between 'societal expectations, environmental preservation, and economic growth', and became critical of the Centre when this was not granted (Soren 2020, cited in Alam 2020). However, his actions were widely denounced by the Jharkhand Janadhikar Mahasabha, a state-level coalition of peoples' organizations, which called for mass protests against the government. Presently, the Mahasabha decried the JMM-led government's effective support of the Centre and rejected the claim that any such mining investments would work in the interests of an *atmanirbhar* (self-reliant) Jharkhand, or for the welfare of Adivasis, pointing to the realities of land grab.

In Chhattisgarh's Korba district, the CIL subsidiary, South Eastern Coalfields Limited (SECL), is a powerful PSU that projects a professional veneer onto messy and conflictual land acquisition. An SECL official claimed that land acquisition was peaceful, conducted without intermediaries, a means commonly favoured by private companies, but using 'young professional village mobilizers' who talked to people directly and discussed compensation and employment issues. According to an official, 'For approximately 2,000 acres of land, 1,000 jobs were offered … despite people being unskilled, we induct them and give them salaries as high as ₹30,000 per month.'[5] At the same time, activists in Korba alleged that there remained many problems with land acquisition, compensation rates were arbitrary, and women were excluded from employment. Even as SECL officials emphasized that they worked in decidedly more peaceful and fairer ways than their private-sector counterparts, many junior members of the district administration and activists objected to the SECL's lack of accountability. Several believed that SECL was driven by higher-level political collusions and that it did not always cooperate with district authorities.

In Raigarh district in Chhattisgarh, Jindal Power Limited, a key private actor within the mining sector, has set up a coal-based thermal power plant in Tamnar. It has rapidly gained influence and visibility for filling gaps in critical public infrastructure, from schools to community buildings and hospitals. The police too allegedly receive favours from the company, including the use of company vehicles, as part of a known quid pro quo, earning the town the rather unflattering sobriquet of 'Jindalgarh'. However, Jindal did undergo extremely conflictual land acquisition proceedings initially.

During fieldwork in 2015–2016, activists in Korba and Raigarh described the strong pushback from the Chhattisgarh state apparatus, which was in favour of expediting clearances to allow companies to extract more quickly and without interruption. In Korba, local forest officials were reluctant to refuse forest clearances for fear of reprisals from higher-ups in the government. Referring to the provisions of the Panchayat (Extension to Scheduled Areas) Act (PESA), 1996, a senior revenue official bluntly said, 'Ultimately, you have to take the coal – these issues (meaning laws) are just obstacles.'[6] Together with the Forest Rights Act, 2006, PESA is an important instrument in the fight against the environmental impacts of extractive development, but it remains underutilized.

Both Korba and Raigarh are environmentally sensitive. Korba is part of the Hasdeo Arand reserve and one of the largest coal reserves in the country. In 2010, it became the subject of an intense debate between the then environment minister, Jairam Ramesh, who sought to declare the forest a 'no-go' area, and the coal ministry, in addition to the Chhattisgarh government itself; Ramesh lost the fight (*The Hindu* 2011). Adivasi groups have contested the government's 2014 decision to allow commercial mining and auction coal blocks (Choudhury 2018). In June 2020, when three coal blocks in Hasdeo Arand, including two in Korba, were included in the list of 41 coal blocks to be auctioned by the Modi government, there was public outcry,

---

[5] Field interview, Korba, April 2015.

[6] Field interview, Korba, April 2015.

most of all from local environmental organizations, like the Manthan Adhyayan Kendra and the Chhattisgarh Bachao Andolan. The Congress-led government of Chhattisgarh joined the protests against the central government, which Jharkhand spearheaded (Jamwal 2020). The Centre responded by dropping the blocks in Hasdeo Arand but included three new mines in the Raigarh district. This move was also widely criticized, as Raigarh is considered a 'toxic hot spot', reeling from the long-term polluting effects of mining activities in and around Tamnar (Khan 2020).

These examples illustrate that Chhattisgarh has a network of activists soldiering on against coal mining. Fieldwork also revealed some differences in the intensity of resistance against the public SECL versus the private Jindal – the latter was considerably more charged. Where private companies are concerned, the already watered down notion of 'public interest', widely evoked in state-led acquisition processes, takes a further knocking (Levien 2013). Activists interviewed in Raigarh provided vivid accounts of the shooting of a prominent local activist, which even led to unverified allegations of a company-sponsored contract killing. Public-sector companies like SECL faced resistance, too, but it was admittedly much tamer; in Korba, activists mainly took to filing public interest litigation (PIL) and Right to Information (RTI) requests. Moreover, as a company representative put it, these protests 'had not caused any disruption'![7]

The irony was that according to SECL, Korba, one of three districts in north Chhattisgarh where the Hasdeo Arand forest is situated, and which has seen sustained Adivasi opposition, was one of the 'easiest places in the country in which to acquire land'.[8] Importantly, neither party had a local elected representative who was willing to support any act of resistance; indeed, a lone BJP politician in Korba was marginalized within his party for raising issues of compensation and employment. These dynamics attest to conditions in the broader extractive regime of Chhattisgarh, where a unified pro-extraction political discourse, combined with a relatively cohesive political command over the institutional apparatus, enables forceful state pushback against resistance. In Jharkhand, the extractive regime is marked by a complicated public, political position on mining issues versus Adivasi lands and rights. While its highly inept state machinery has become better at targeting protestors – especially under the BJP's rule from 2014 to 2019 – the extractive regime here still needs to contend with the plural, vibrant forces of Adivasi political resistance. Importantly, as this chapter has discussed, activism in Jharkhand often has the support of Adivasi political parties, a critical variable that is missing in Chhattisgarh.

[7] Field Interview, Korba, April 2015.
[8] Field Interview, Korba, April 2015.

Analysis: extractive regimes and implications for a just transition

The preceding analysis demonstrates that understanding the extractive regimes of these coal-producing states is necessary to grasp the prospects of bargaining and resistance in ongoing and new frontiers of coal mining. In this section, I turn to the implications of extractive regimes for the prospects of a just energy transition.

Coal communities that bear the historical brunt of extraction also risk re-victimization during an energy transition. If coal mines are rendered unprofitable and close down, as is already happening in Ramgarh district in Jharkhand, then there will be direct implications for workers' livelihoods (Bhushan, Banerjee, Agarawal 2020). Macro-analysts are circumspect about the overall labour impacts of the coal transition on the grounds and estimate that employment creation/decline in the coal industry will range from +79,000 to –40,000 by 2030 compared to today, whereas coal-rich states will produce around 45 million new entrants to the labour market by 2030 (Spencer et al. 2018). In other words, we cannot look to the coal sector to accommodate the labour needs of the coal heartlands. However, contrary to the notion that extractive regimes would, therefore, matter less, this chapter argues that they continue to be deeply significant to what lies ahead.

Soren recently said, 'We are mindful that coal will reduce over time, and therefore, we have to plan for a *post-coal* future. As Jharkhand is rich in other natural resources, we are diversifying our economy and promoting tourism, forest, agro-based industries, and the service sectors' (IANS 2020, emphasis mine). Chhattisgarh's state-run power distribution company announced in 2019 that it will not build any more coal-fired power plants; NTPC's 1,600 megawatt (MW) power plant in Raigarh would meet half of the state's needs and be the last such plant in the state (Rathi and Singh 2019). Besides, both states are committed to investing in renewables, especially large-scale solar power plants, as part of their perceived sub-national duty to contribute to ambitious national goals (Mazumdar 2015). In line with this strategy, in 2019, Chhattisgarh allocated almost 400 hectares of land for a solar project in Rajnandgaon district.

Scholars report chilling similarities in the dynamics of land acquisition and dispossession in RE projects and coal mining, although their distinctive modalities merit systematic examination. Solar projects are deemed more suitable for the coal heartlands as compared with wind (Pai et al. 2020). However, solar projects are responsible for the creation of a precariat, given the largely 'jobless growth' of solar projects elsewhere, which offer far fewer prospects for formal employment compared to the CIL during its heyday. The spread of coercive and extra-legal practices such as enclosure, land grab, and divestment of the commons has been noted in 'green' energy–related land acquisitions across India (Stock and Birkenholtz 2019; Yenneti, Day, and Golubchikov 2016). Moreover, the alliance of state and corporate interests that drives solar and wind projects in India 'downplays the narratives of dispossession, treating people and their livelihoods resources as worthy of sacrifice for the sake of societally beneficial green energy transitions' (Yenneti, Day, and Golubchikov 2016, 13). Poor people, who are most dependent on the commons, will disproportionately bear the cost of RE development, just like in the case of coal mining. The governance of large solar projects in the coal heartlands is still not well understood and, indeed, presents a new area for urgent, critical research.

While both states are courting private investors for RE and facilitate land acquisition, Jharkhand will almost certainly engage more widely with political representatives and activists than Chhattisgarh. There may be greater tolerance of dissent in Jharkhand, just as with coal mining so far. Extractive regimes at the sub-national level will play a massive part in determining the political character of any transition away from coal and its prospects of fairness. Whether such a transition is bottom-up or top-down will follow on from extant regimes. The hallmark of a bottom-up transition is the creation of a political space for broad-based consultations with actors and networks advocating for the rights of victimized groups.

Indeed, to achieve a just transition, there needs to be a broader societal dialogue on labour; indeed, Western countries like Germany and Poland have attempted to do this in their transitions.[9] As not all coal workers in India are unionized, or even informally organized, engaging with civil society networks – like the JMACC in Jharkhand and the CBA in Chhattisgarh, among many others – becomes essential. The networks' historical sensitivity to and perspectives on changing dynamics in local resource dependence and livelihoods make them invaluable partners in steering the transition. While it is hard to imagine the successful formation of such partnerships in the upper-caste-dominated state-corporate alliance in Chhattisgarh, there is still hope for progressive spaces and opportunity in Jharkhand.

But there is a need for caution. Jharkhand's extractive regime is notorious for bureaucratic apathy and proclivity for fragmentation and rent-seeking, which have historically impeded the formulation of coherent policies around resettlement and compensation. These are likely to remain problems. Another important possibility is that neither state, despite significant differences in their extractive regimes, will effectively challenge corporate practices, either public or private. The fieldwork vignettes presented earlier confirm that coal companies exercise considerable

---

[9] A spokesperson for the European Trade Union Confederation made this point at a webinar discussing the 'Just transition in India' on 9 December 2020.

political sway, either through collusion with the local state, or even against its wishes, when higher-level political support exists. These dynamics are unlikely to change with RE projects; somewhat predictably, given the 'big capital' nature of investments, many of the same private actors leading the privatization of coal in 'new geographies' are also heavily invested in wind and solar projects (Oskarsson et al. 2021). Recent research also suggests that future CIL reconfigurations, driven by the anticipated decline in coal consumption by thermal power plants, will involve new forms of greenfield development, potential land acquisition, and public–private partnerships with domestic and foreign actors (Rajshekhar 2021).

There are also important transformations underway, in terms of both pacification and protests across India, and these will shape the prospects of just transitions in both extractive regimes. With privatization, decreased labour intensity, and greater labour informalization, both coal and RE projects will need land and not labour, requiring a reconfiguring of relationships with local communities. Kale (2020, 1213) argues that private companies are increasingly relying on corporate social responsibility (CSR) to manage, discipline, and pacify local communities. Protest movements also reflect Adivasis' increased discontent with compensation and jobs and, indeed, 'compensatory jobs for dispossession' have become a source of stratification within Adivasi communities (Noy 2020, 388). There are, however, worrying signs that NGOs (non-governmental organizations) and protest movements that limit themselves to demanding better land prices or settlement packages are the ones that are likely to survive the might of a repressive extractive regime (Das Gupta [2019] reports on this in Raipur, Chhattisgarh, and there are no doubt others). One of the greatest issues associated with a just transition is whether there will be any space to question the highly inequitable modes of capitalist expansion that are reconfiguring the frontiers of extraction.

Conclusion

This chapter addresses difficult questions concerning a just transition when viewed from the vantage point of the coal heartlands. Communities at the coal frontiers have borne the nested injustices of coal mining and energy poverty and now face an uncertain future given the overall future of coal in India. The chapter demonstrates that current strategies for facilitating a just transition from coal, like retraining workers or filling gaps in employment through RE development, are extremely limited. Energy transitions involve the reconfiguration of social practices and political power. By drawing attention to extractive regimes, which clearly drive injustices and limit prospects for fairness in coal-mining areas, and by arguing that the switch to RE projects needs to be viewed as a continuum that unfolds within the same extractive regimes, this chapter makes a potentially novel contribution to the growing critical scholarship on just transitions. Looking ahead, given the increased privatization of extractive industries, the role of sub-national states is only set to grow in importance. There will be ever new temptations to attract investments, both domestic and foreign, and further imperatives for oppression. While both extractive regimes in the coal heartlands may be increasingly intolerant of resistance, with some important differences between them, just transition advocates must demand bottom-up political engagement with the multiple actors and networks that agitate for justice in coal mining today. Indeed, this may be the only hope in resisting novel frontiers of injustice among historically disadvantaged lands and their peoples.

Acknowledgements

This chapter is based on research funded by the Effective States and Inclusive Development Research Centre, University of Manchester, 2014–2017. I am grateful to Anindita Adhikari and Vidushi Bahuguna for their valuable intellectual partnership during this research. I thank Patrik Oskarsson, Anindita Adhikari, Aditya Valiathan Pillai, David Singh, and the anonymous reviewers of this edited volume for their constructive comments on previous drafts. Particular thanks are due to the volume editor, Prakash Kashwan, for his incisive observations on this chapter.

References

Adhikari, A. and V. Chhotray, 2020. 'The Political Construction of Extractive Regimes in Two Newly Created Indian States: A Comparative Analysis of Jharkhand and Chhattisgarh'. *Development and Change* 51 (3): 843–873. https://doi.org/10.1111/dech.12583.

Alam, M. 2020. 'Jharkhand-based coalition calls for mass protests against decision to open coal sector'. *Down to Earth*, 18 June. https://www.downtoearth.org.in/news/mining/jharkhand-based-coalition-calls-for-mass-protests-against-decision-to-open-coal-sector-71847 (accessed 23 June 2020).

Arsel, M., B. Hogenboom, and L. Pellegrini. 2016. 'The Extractive Imperative in Latin America'. The Extractive Industries and Society 3 (4): 880–887. https://doi.org/10.1016/j.exis.2016.10.014.

Banerjee, S. 2020. 'District Mineral Foundation Funds Crucial Resource for Ensuring Income Security in Mining Areas Post COVID-19'. *Brookings*, 6 May. https://www.brookings.edu/blog/up-front/2020/05/06/district-mineral-foundation-funds-crucial-resource-for-ensuring-income-security-in-mining-areas-post-covid-19/ (accessed 20 May 2020).

Barik, S. 2020. 'Hemant Soren Says Centre's Decision on Coal Mine Auction "a Disregard of Cooperative Federalism"'. *The Hindu*, 20 June. https://www.thehindu.com/news/national/

hemant-soren-says-centres-decision-on-coal-mine-auction-a-disregard-of-cooperative-federalism/article31880935.ece (accessed 26 June 2020).

Berthet, S. 2011. 'Chhattisgarh: Redefining the Role of the State?' *In New States for a New India: Federalism and Decentralisation in the States of Jharkhand and Chhattisgarh*, edited by S. Berthet and G. Kumar. New Delhi: Manohar Publishers, 87–115.

Bhattacharjee, S., 2017. India's Coal Story: *From Damodar to Zambezi*. New Delhi: SAGE Publishing.

Bhushan, C., S. Banerjee, and S. Agarawal. 2020. *Just Transition in India: An Inquiry into the Challenges and Opportunities for a Post-coal Future*. New Delhi: iFOREST.

Bidwai, P. 2012. *The Politics of Climate Change and the Global Crisis: Mortgaging Our Future*. New Delhi: Orient Blackswan.

Bridge, G. and L. Gailing. 2020. New Energy Spaces: Towards a Geographical Political Economy of Energy Transition. *Environ Plan* A 52 (6): 1037–1050. https://doi.org/10.1177/0308518X20939570.

Chakraborty, L., S. Garg, and G. Singh. 2016. 'Cashing in on Mining'. WP 16/161, NIPFP. http://www.academia.edu/download/42803966/WP_2016_161.pdf (accessed 3 April 2020).

Chakravarty, S. and M. V. Ramana. 2012. 'The Hiding behind the Poor Debate: A Synthetic Overview'. *In Handbook of Climate Change and India: Development, Politics and Governance*, edited by N. Dubash., New Delhi: Routledge, 218–230.

Chandra, R. 2018. 'Adaptive State Capitalism in the Indian Coal Industry'. Ph.D. Dissertation, Harvard University.

Chatterjee, E., 2020. 'The Asian Anthropocene: Electricity and Fossil Developmentalism'. *Journal of Asian Studies* 79 (1): 3–24. https://doi.org/10.1017/S0021911819000573.

Choudhury, C. 2018. 'Taking Over Fertile Land for Adani Group from Protesting Farmers'. *India Spend*, 1 December. https://www.indiaspend.com/taking-over-fertile-land-for-adani-group-from-protesting-farmers-jharkhand-government-manipulates-new-law-meant-to-protect-them/ (accessed 5 April 2020).

CNBC-TV18. 2019. 'Chhattisgarh Government Returned Land Acquired by the Tata Group in Bastar'. 7 May. https://www.cnbctv18.com/politics/chhattisgarh-government-returned-land-acquired-by-the-tata-in-bastar-to-the-tribals-says-rahul-gandhi-in-chaibasa-3229341.htm (accessed 7 April 2020).

CSE. 2008. *Rich Lands, Poor People: Is Sustainable Mining Possible? State of India's Environment: A Citizens' Report*, No. 6, New Delhi: Centre for Science and Environment (CSE).

Das Gupta, S. 2019. *Class, Politics, and Agrarian Policies in Post-liberalisation India*. Cambridge: Cambridge University Press. https://doi.org/10.1017/9781108236201.

D'Costa, A. P. and A. Chakraborty. 2017. *The Land Question in India: State, Dispossession, and Capitalist Transition*. Oxford: Oxford University Press.

Edwards, G. A. S. 2019. 'Coal and Climate Change'. *WIREs Climate Change* 10 (5). https://doi.org/10.1002/wcc.607.

Hebbar, R. 2003. 'From Resistance to Governance'. *Seminar*. https://www.india-seminar.com/2003/524/524%20ritambhara%20hebbar.htm, accessed 16 August 2020.

IANS. 2020. 'Energy Transition Must Go Hand in Hand with "Just Transition": Study'. *Economic Times*, 25 November. https://energy.economictimes.indiatimes.com/news/coal/energy-

transition-must-go-hand-in-hand-with-just-transition-study/79400829,   accessed   12
     December 2020.

Jaitly, A. 2021. 'Excuse Me, Mr U.N. Secretary-General'. *Bloomberg*, 5 March. https://www.
     bloombergquint.com/opinion/coal-power-plants-and-climate-change-excuse-me-mr-un-
     secretary-general (accessed 12 April 2021).

Jamwal, N. 2020. 'After Jharkhand, Chhattisgarh and Maharashtra Oppose the Centre's Auction'.
     Gaon Connection, 15 June. https://en.gaonconnection.com/after-jharkhand-chhattisgarh-
     and-maharashtra-oppose-the-centres-auction-of-41-coal-blocks-for-commercial-mining/
     (accessed 12 August 2020).

Kale, S. S. 2020. 'From Company Town to Company Village: CSR and the Management of Rural
     Aspirations in Eastern India's Extractive Economies'. *Journal of Peasant Studies* 47 (6):
     1211–1232. https://doi.org/10.1080/03066150.2020.1825290.

Kale, S. S. and N. Mazaheri. 2014. 'Natural Resources, Development Strategies, and Lower Caste
     Empowerment in India's Mineral Belt: Bihar and Odisha During the 1990s'. *Studies in
     Comparative International Development* 49 (3): 343–369. https://doi.org/10.1007/s12116-
     014-9162-2.

Kashwan, P. 2017. *Democracy in the Woods: Environmental Conservation and Social Justice in
     India, Tanzania, and Mexico*. New York: Oxford University Press.

Khan, S. 2020. 'Mine Games'. *Gaon Connection*, 15 June. https://en.gaonconnection.com/mine-
     games-five-contentious-coal-blocks-in-chhattisgarh-dropped-from-the-auction-list-but-
     three-objectionable-mines-added/ (accessed 11 August 2020).

Kumar, S. 2018. 'Adivasis and the State Politics in Jharkhand'. *Studies in Indian Politics* 6 (1):
     103–116, https://doi.org/10.1177/2321023018762821.

Lahiri-Dutt, K. 2014. *The Coal Nation: Histories, Ecologies and Politics of Coal in India*. Surrey:
     Routledge.

———. 2016. 'The Diverse Worlds of Coal in India: Energising the Nation, Energising
     Livelihoods'. *Energy Policy* 99: 203–213. https://doi.org/10.1016/j.enpol.2016.05.045.

Lakhanpal, S. 2019. 'Contesting Renewable Energy in the Global South: A Case-Study of Local
     Opposition to a Wind Power Project in the Western Ghats of India'. *Environmental
     Development* 30 (June): 51–60. https://doi.org/10.1016/j.envdev.2019.02.002.

Lazarus, M. and H. van Asselt. 2018. 'Fossil Fuel Supply and Climate Policy: Exploring the Road
     Less Taken'. *Climatic Change* 150 (1–2): 1–13. https://doi.org/10.1007/s10584-018-2266-3.

Levien, M. 2013. 'The Politics of Dispossession: Theorizing India's "Land Wars".' *Politics and
     Society* 41 (3): 351–394. https://doi.org/10.1177/0032329213493751.

———. 2018. *Dispossession without Development: Land Grabs in Neoliberal India*. New York:
     Oxford University Press. https://doi.org/10.1080/03066150.2012.671768.

Mahapatra, D. 2020. 'SC Criticises Jharkhand's doublespeak'. *Energyworld.com*, 1 October. https://
     energy.economictimes.indiatimes.com/news/coal/sc-criticises-jharkhands-doublespeak-
     on-coal-mine-auction/78423264 (accessed 5 December 2020).

Mazumdar, G. 2015. 'Jharkhand Gets Ready for Big Leap in Solar Power'. *Hindustan Times*, 31
     July. https://www.hindustantimes.com/ranchi/jharkhand-gets-ready-for-big-leap-in-solar-
     power/story-SUVBVG1ImLWkwzZK00xqiP.html (accessed 10 August 2020).

Newell, P. 2019. 'Trasformismo or Transformation? The Global Political Economy of Energy Transitions'. *Review of International Political Economy* 26 (1): 25–48. https://doi.org/10.1 080/09692290.2018.1511448.

Newell, P., and D. Mulvaney. 2013. 'The Political Economy of the "Just Transition"'. *Geographical Journal* 179 (2): 132–140.

Noy, I. 2020. 'Public Sector Employment, Class Mobility, and Differentiation in a Tribal Coal Mining Village in India'. *Contemporary South Asia* 28 (3): 374–391. https://doi.org/10.1080 /09584935.2020.1801579.

Oskarsson, P., K. Lahiri-Dutt, and P. Wennström. 2019. 'From Incremental Dispossession to a Cumulative Land Grab: Understanding Territorial Transformation in India's North Karanpura Coalfield'. *Development and Change* 50 (6): 1485–1508. https://doi.org/10.1111/ dech.12513.

Oskarsson, P., K. B. Nielsen, K. Lahiri-Dutt, and B. Roy. 2021. 'India's New Coal Geography: Coastal Transformations, Imported Fuel and State-Business Collaboration in the Transition to More Fossil Fuel Energy'. *Energy Research and Social Science* 73: 1–10. https://doi. org/10.1016/j.erss.2020.101903.

Pai, S. 2018. *Total Transition: The Human Side of the Renewable Energy Revolution.* British Columbia: Rocky Mountain Books Incorporated.

Pai, S., H. Zerriffi, J. Jewell, and J. Pathak. 2020. Solar Has Greater Techno-Economic Resource Suitability than Wind for Replacing Coal Mining Jobs'. *Environmental Research Letters* 15 (3): 1–13. https://doi.org/10.1088/1748-9326/ab6c6d.

Pai, S., K. Harrison, H. Zerriffi. 2020. 'A Systematic Review of the Key Elements of a Just Transition for Fossil Fuel Workers'. WP 20-04, Clean Economy Working Paper Series, Smart Prosperity Institute.

Rajalakshmi, T. K. 2016. 'Assault on Tribal Rights'. *Frontline*, 25 November. https://frontline. thehindu.com/the-nation/assault-on-tribal-rights/article9319983.ece (accessed 10 April 2020).

Rajshekhar, M. 2012. 'Chhattisgarh Power Boom That Never Was'. *Economic Times*, 25 October. https://m.economictimes.com/industry/energy/power/chhattisgarh-power-boom-that-never-was-only-15-out-60-thermal-plants-may-get-operational/articleshow/16946524.cms (accessed 10 August 2020).

———. 2020. 'Why India's GHG emissions Are about to Rise Faster'. *Carboncopy*, 30 June. https://carboncopy.info/why-indias-ghg-emissions-are-about-to-rise-faster/ (accessed 11 August 2020).

———. 2021. 'Strange Times Ahead for India's Coal Sector'. *Carboncopy*, 2 April. https:// carboncopry.info/strange-times-ahead-for-indias-coal-sector/ (accessed 7 April 2021).

Rathi, A. and K. Singh. 2019. 'After Gujarat India's Chhattisgarh Won't Build Coal Power Plants'. *Quartz India*, 16 September. https://qz.com/india/1709483/after-gujarat-indias-chhattisgarh-wont-build-coal-power-plants/ (accessed 9 April 2020).

Sathe, D. 2016. 'Land Acquisition: Need for a Shift in Discourse?' *Frontline* 51: 52–58.

Shah, A. 2006. 'Markets of Protection: The 'Terrorist' Maoist Movement and the State in Jharkhand, India'. *Critique of Anthropology* 26 (3): 297–314. https://doi.org/10.1177/0308275X06066576.

Sharma, S. 2019. 'Litmus Test for Congress in Chhattisgarh'. *Scroll.in*, 26 March. https://scroll.in/ article/917827/litmus-test-for-congress-in-chhattisgarh-will-it-hand-over-another-coal-mine-to-the-adani-group (accessed 11 April 2020).

Singh, N. and B. Harriss-White. 2019. 'The Criminal Economics and Politics of Black Coal in Jharkhand, 2014'. *In The Wild East: Criminal Political Economies in South Asia*, edited by B. Harriss-White and L.Michelutt. London: UCL Press, 35–67.

Spencer, T., R. Pachauri, G. Renjith, and S. Vohra. 2018. 'Coal Transition in India'. Discussion Paper, TERI.

Sra, G. 2020. 'Coal Mining and Community Activism in India'. *The Ecologist*, 21 January. https://theecologist.org/2020/jan/21/coal-mining-and-community-activism-india.

Stock, R. and T. Birkenholtz. 2019. 'The Sun and the Scythe: Energy Dispossessions and the Agrarian Question of Labor in Solar Parks'. *Journal of Peasant Studies* 48 (5): 984–1007. https://doi.org/10.1080/03066150.2019.1683002.

Sud, N. 2019. 'The Unfixed State of Unfixed Land'. *Development and Change* 51 (5): 1175–1198. https://doi.org/10.1111/dech.12553.

Sundar, N. 1997. *Subalterns and Sovereigns: An Anthropological History of Bastar, 1854–1996.* Oxford: Oxford University Press.

———. 2007. *Legal Grounds: Natural Resources, Identity and the Law in Jharkhand.* New Delhi: Oxford University Press.

Thachil, T. 2014. *Elite Parties, Poor Voters: How Social Services Win Votes in India.* Cambridge Studies in Comparative Politics, Cambridge: Cambridge University Press, https://doi.org/10.1017/CBO9781107707184.

Tillin, L. 2013. *Remapping India: New States and Their Political Origins.* Oxford: Oxford University Press.

The Hindu. 2011. 'Jairam Loses "No-Go" Battle'. 24 June. https://www.thehindu.com/news/national/jairam-loses-nogo-battle-allows-coal-mining-in-forested-hasdeo-arand/article2129774.ece (accessed 7 April 2020).

TNN. 2020. 'Jharkhand Will Set Up Panel to Return Acquired but Unused Land'. *Times of India*, 13 March. https://timesofindia.indiatimes.com/city/ranchi/will-set-up-panel-to-return-acquired-but-unused-land-jharkhand-cm/articleshow/74610671.cms (accessed 31 March 2020).

Tongia, R. and S. Gross. 2019. 'Coal in India: Adjusting to transition'. WP 7, Brooking India.

Yadav, A. 2013. 'Why the Land Wars Won't End'. *The Hindu*, 29 September. https://www.thehindu.com/opinion/op-ed/why-the-land-wars-wont-end/article5182473.ece (accessed 19 April 2020).

Yenneti, K., R. Day, and O. Golubchikov. 2016. 'Spatial Justice and the Land Politics of Renewables: Dispossessing Vulnerable Communities through Solar Energy Mega-projects'. *Geoforum* 76: 90–99. https://doi.org/10.1016/j.geoforum.2016.09.004.

Wahi, N. and A. Bhatia. 2018. 'The Legal Regime and Political Economy of Scheduled Tribes in Scheduled Areas of India'. https://papers.ssrn.com/sol3/papers.cfm?abstract_id=3759219 (accessed 1 December 2020).

Zinecker, W. A., P. Gass, I. Gerasimchuk, P. Jain, and T. Moerenhout. 2014. 'Real People, Real Change – Strategies for Just Energy Transitions'. International Institute for Sustainable Development.

Green Solidarity by Anupriya

Climate Justice Implications of the Relationship between Economic Inequality and Carbon Emissions in India

Haimanti Bhattacharya

Introduction

The carbon emissions of the world's richest 1 per cent are more than double the emissions of the poorest 50 per cent, despite the fact that climate change is expected to disproportionately affect the poor, especially in the warmer parts of the world (Oxfam 2020; Goswami 2020). This suggests that climate and socio-economic justice are intertwined, and understanding the nature of the relationship between carbon emissions and economic inequality can help us arrive at potential pathways to address both.

Climate and socio-economic justice are crucial in the case of India. India is a significant player in the global economy, as its gross domestic product (GDP) is the fifth largest in the world (World Bank 2021). However, the country also experiences staggering levels of economic inequality. It has the third highest number of billionaires, but it also has the largest poor population in the world (Ankel 2020; Roser and Ortiz-Ospina 2019). The wealth of the richest 1 per cent of the Indian population is more than four times the total wealth of the bottom 70 per cent (*Economic Times* 2020). These indicators of economic inequality demonstrate a dire need to enhance socio-economic justice in India.

Three different indicators of carbon emissions are widely used for analytical purposes. The present climate crisis resulted from historically accumulated greenhouse gas emissions, measured in carbon dioxide equivalents. The United States (US) is the largest contributor, accounting for about 25 per cent of the global cumulative carbon emissions between 1751 and 2019, while India contributed about 3 per cent (Ritchie 2019). Hence, India's historical cumulative carbon emission is rather low. The second indicator measures annual carbon emissions, or a country's current emission levels. Based on this indicator, India has the third highest carbon emission levels globally; it trails China and the US by a huge margin.[1] The third indicator is per capita annual emissions, which accounts for differences in the population size of countries. When countries are ranked in descending order of their per capita carbon emissions, India ranked 128 out of 210 countries in 2019 (Crippa et al. 2020). Even though India's per capita carbon emissions and its share in cumulative global emissions is low, its current scale of emissions is a matter of concern (Matthews 2016). Therefore, this analysis focuses on the scale of annual carbon emissions in India.

India's carbon emissions are concerning because despite its low per capita emissions, many Indian cities experience high levels of air pollution. Thirty-five out of the 50 most polluted cities globally are located in India (IQAir 2020). The effects of air pollution are most heavily borne by India's poorest people, who lack relevant protection both at the workplace and at home. Furthermore, since India is located in a warm region and has the largest number of poor globally, the poor in India will be exposed to a disproportionate share of climate change impacts. Technologies that help reduce dependence on fossil fuels, and thereby reduce carbon emissions, face serious obstacles in India, as the high concentration of economic resources in the hands of a few in the top economic strata leave the vast majority of the population with meagre resources to adopt such carbon mitigation technologies (Gill 2021). Therefore, the relationship between carbon emissions and economic inequality has important implications for climate and socio-economic justice in India. And since India has the second largest population globally, socio-economic and climate justice in the country also have consequences for the rest of the globalized world.

This chapter's core argument is that the relationship between annual carbon emissions and economic inequality has undergone a fundamental transformation in the post-liberalization period in India. My research shows that in the pre-liberalization period, economic inequality at the state level had a negative association

---

[1] In 2020, China's carbon emission was 4.4 times that of India, and the US emitted about twice that of India.

with carbon emissions. However, in the years following the introduction of a wide range of economic liberalization measures, higher economic inequality at the state level came to be associated with higher carbon emissions. This finding resonates with evidence of a positive relationship between carbon emissions and economic inequality in the US and China. This suggests that like the top two emitters, India has an opportunity to adopt a holistic approach towards economic development that mitigates carbon emissions and economic inequality jointly, thereby fostering climate and socio-economic justice.

The rest of the chapter is organized as follows. The following section summarizes the insights from the existing literature on the relationship between carbon emissions and economic inequality. The third section describes the data and methodology of my research. The fourth section discusses the key results. The fifth section discusses the key implications and the final section offers a conclusion.

Background and past research

The empirical literature on income distribution or income inequality as a driver of carbon emissions is mainly based on international data. Among the international studies, Grunewald et al. (2017) provide the most comprehensive spatial and temporal coverage, with data from 158 countries for the period 1980–2008. Grunewald et al. (2017) find that while higher inequality is associated with lower per capita emissions in lower-income countries, higher inequality is associated with higher per capita emissions in higher-income countries.

Evidence on how economic inequality influences annual carbon emissions at the intra-country level is relatively sparse. The literature provides intra-country evidence of economic inequality as a driver of carbon emissions for the top two carbon-emitting countries: the US and China. Jorgenson, Schor, and Huang (2017) examined state-level data from the US for 1997–2012 and found that a higher concentration of income among the wealthiest 10 per cent of the population is associated with higher levels of carbon emissions. Using a subnational regional-level panel dataset from China for the period 1995–2010, Zhang and Zhao (2014) found a qualitatively similar result – higher income inequality is associated with higher carbon emissions.

There remains a paucity of evidence on the relationship between the scale of carbon emissions and economic inequality at the subnational level. This creates a major knowledge gap for policymaking in countries with federalism, as policymaking powers and enforcement are divided between national and subnational governments.

In these countries, economic policies are likely to be designed and implemented at the subnational level.

This chapter builds on research published by Bhattacharya (2020), which used state-level panel data from India for 1981–2008. This study examines the influence of a major policy change – India's 1991 economic liberalization policy – on the relationship between state-level carbon emissions and economic inequality. The results demonstrate that policy changes can completely alter the relationship between carbon emissions and economic inequality. The nature of this relationship has important implications for national and subnational climate action and socio-economic justice for India's poorest citizens.

Data and methodology

The empirical analysis focuses on the relationship between carbon emissions and economic inequality based on data from 14 major states in India for the period 1981–2008. These states are Andhra Pradesh (AP), Assam (AS), Bihar (BR), Gujarat (GJ), Karnataka (KA), Kerala (KL), Madhya Pradesh (MP), Maharashtra (MH), Orissa (OR), Punjab (PB), Rajasthan (RJ), Tamil Nadu (TN), Uttar Pradesh (UP), and West Bengal (WB). These states cover a vast majority of India's geographic area (see Figure 5.1) and are also salient in terms of size of economy and population – factors that no doubt influenced data availability for these states.[2] The data sources and descriptions are provided in Table 5.1.

I used the state-level anthropogenic carbon emissions based on fossil fuel use estimated by Ghoshal and Bhattacharya (2008, 2012) as the outcome variable for this analysis. In India and globally, the largest source of anthropogenic carbon emissions is fossil fuel use (Garg and Shukla, 2002; Ghoshal and Bhattacharya, 2012). I did not include other sources of anthropogenic carbon emissions, like deforestation, land-use changes, soil erosion, and agriculture, due to a lack of adequate data.

I used the state-level Gini index of inequality in consumption expenditure, as estimated by Das, Sinha, and Mitra (2014), as an indicator of economic inequality that is the explanatory variable of primary interest. Economic inequality is usually measured in terms of income inequality, as income encapsulates both consumption expenditure and savings that generate wealth. However, in India, reliable, disaggregated income data is difficult to procure, as over 80 per cent of the workforce is employed in the unorganized informal sector (*The Wire* 2018). Therefore, consumption expenditure data collected by the National Sample Survey Organization is often used to construct indicators of economic inequality in India.

---

[2] A major state, Haryana, is not included in the analysis due to lack of data on inequality.

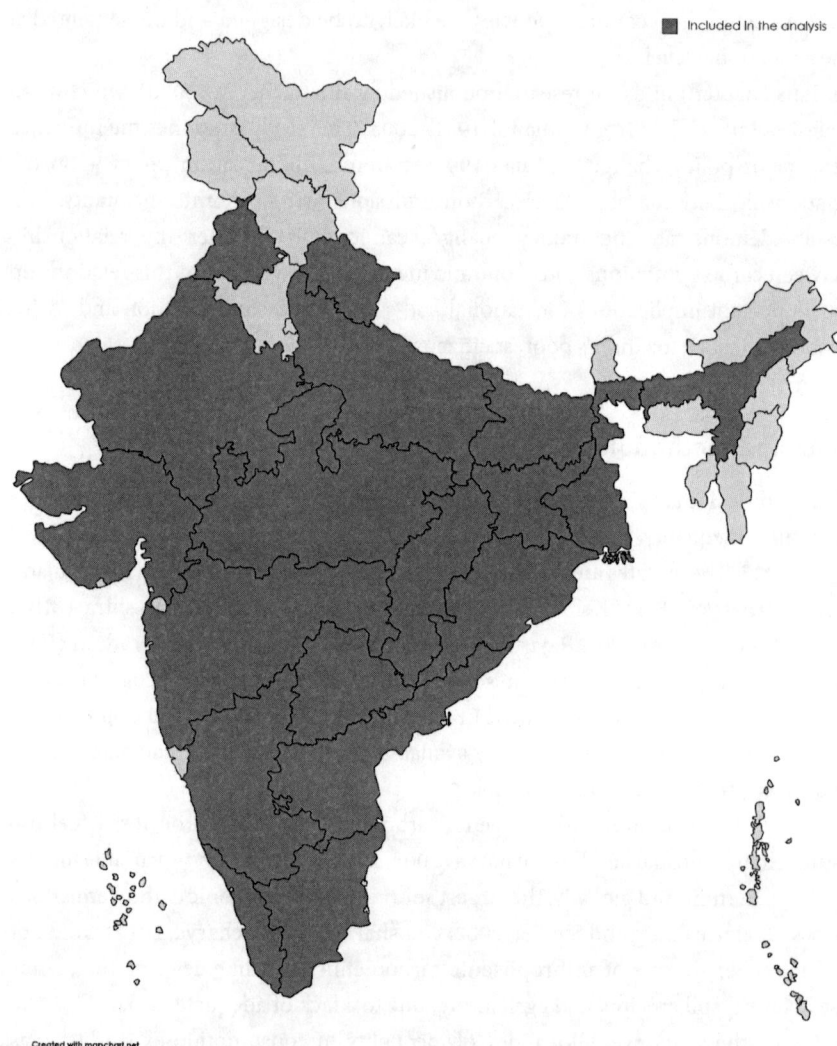

Figure 5.1 Geographical scope (shaded dark) of empirical analysis
Source: Created by author using https://mapchart.net/india.html.
Note: Map not to scale and does not represent authentic international boundaries.

Table 5.2 presents the summary statistics of the variables that I used in my analysis. The measures of variability between states, and the measures of variability within states over time, are substantive for all the variables (see the 'Between Standard Deviation' and 'Within Standard Deviation' columns). Such variability in the data is desirable for multivariate regression analysis.

Table 5.1 Data description

Variable name	Description	Data source
Carbon	Annual state-level carbon-dioxide emissions measured in thousands of metric tons	T. Ghoshal and R. Bhattacharya, 'State-Level Carbon Dioxide Emissions of India: 1980–2000', *Arthaniti – Journal of Economic Theory and Practice 7*, nos. 1–2 (2008): 41–73, and T. Ghoshal and R. Bhattacharyya, 'Carbon Dioxide Emissions of Indian States: An Update', 2012, https://ssrn.com/abstract =2166900 (accessed 20 April 2019).
Gini	Gini coefficient for state-level economic inequality measured on a scale of 0 to 100	S. Das, G. Sinha, and T. K. Mitra, 'Economic Growth and Income Inequality: Examining the Links in Indian Economy', *Journal of Quantitative Economics 12*, no. 1 (2014): 86–95.
GSDP	Gross state domestic product at factor cost at 1980–1981 constant prices measured in billions of rupees	Ministry of Statistics and Programme Implementation, Government of India, http://mospi.nic.in/data.
Industry	The percentage share of mining, manufacturing, electricity, gas, and water supply in state domestic product	
Service	The percentage share of construction, transport, storage and communication, trade, hotels and restaurant, banking and insurance, real estate, ownership of dwellings and business services, public administration, defence and quasi-govt. bodies, and other services in state domestic product	
Population	Total state population measured in millions	Population estimate = NDSP/per capita NSDP
Urban	Urban population as a percentage of the total state population	Census of India – 1981, 1991, 2001. Assumed linear growth rate to estimate the urban population for non-census years.

Table 5.2 Statistical summary of the data

Variable	Obser-vations	Mean	Overall Standard Deviation	Between Standard Deviation	Within Standard Deviation	Mini-mum	Maximum
Carbon	392	16,346.14	13,007.58	10,612.01	8,022.40	622.91	7,6159.89
Gini	392	28.09	4.20	2.99	3.05	16.13	38.24
GSDP	392	170.07	131.78	99.04	90.75	25.16	943.97
Industry	392	17.83	5.97	5.20	3.24	2.35	34.38
Service	392	47.91	9.33	5.26	7.83	29.55	79.06
Population	392	57.47	30.29	29.77	9.61	16.64	174.95
Urban	392	26.09	9.57	9.48	2.81	9.84	47.18

Note: Number of states = 14; number of years = 28.

I used regression analysis to examine how the relationship between carbon emissions and economic inequality evolved over 1981–2008. The literature on drivers of carbon emissions shows that the scale and composition of economic activities and the scale and composition of the human population are significant.[3] Hence, the regression analysis focused on estimating the relationship between carbon emissions and economic inequality, controlled for the scale of the state's economy, the percentage share of different sectors in the state's economy, the size of the state's population, and the share of the urban population in the state.

In these 28 years, 1991 is considered a watershed year, as it marks the emergence of economic liberalization in India. In the pre-liberalization period, India's economic policies constrained market forces due to the restrictions imposed by tariff and non-tariff barriers on trade, restrictions on domestic and foreign private investments, state control of banking and insurance, and public-sector monopolies in several industries. Some economic reforms initiated in the 1980s attempted to reduce these restrictions. However, the 1991 economic liberalization policy that opened the Indian economy to foreign trade and investment is considered the most prominent policy change in India's post-independence history, as it triggered such a substantive increase in India's GDP and trade in the subsequent period that it drew global attention.[4] Figures 5.2 and 5.3 highlight the growth of India's economy (represented by GDP) and trade (represented by imports and exports), respectively, in 1981–2008.

[3] See, for example, Jorgenson, Schor, and Huang (2017); Zhang and Zhao (2014).

[4] See Kotwal, Ramaswami, and Wadhwa (2011) for an overview of the impact of economic liberalization on India's economy.

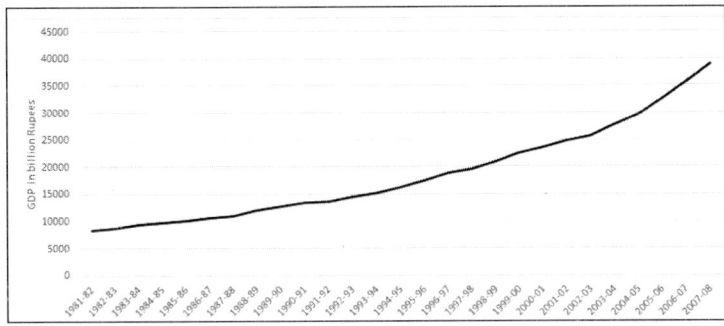

Figure 5.2 GDP of India (at constant 2004–2005 prices)

Data Source: Central Statistical Organization (CSO), India.

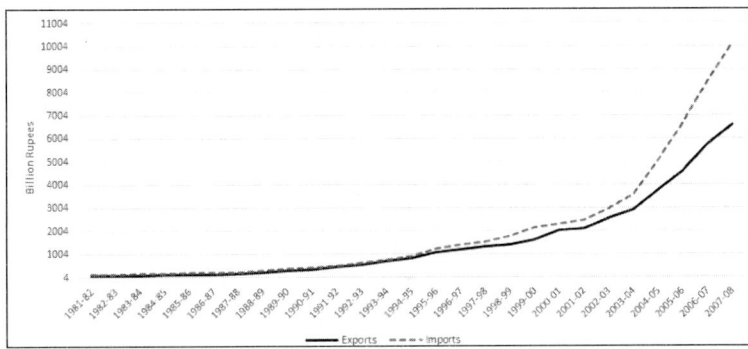

Figure 5.3 Monetary value of exports and imports of India

Data Source: Planning Commission, India.

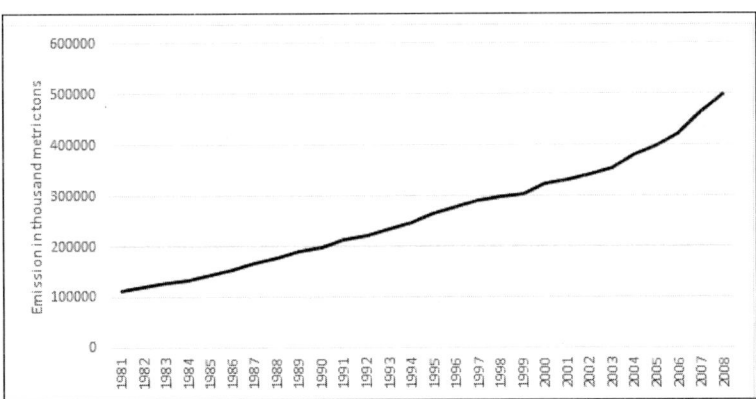

Figure 5.4 Carbon emissions in India

Data Source: Ghoshal and Bhattacharyya (2012).

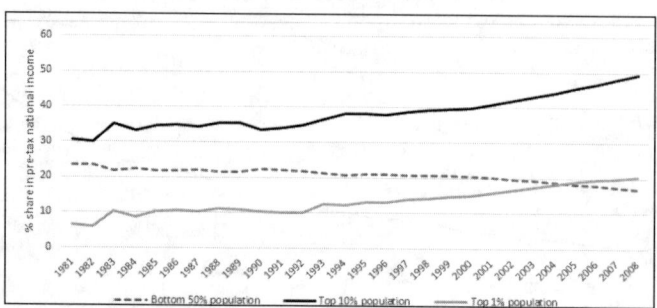

Figure 5.5 Economic inequality indicators for India
Data Source: World Inequality Database (wid.world).

If we look at India's carbon emissions in 1981–2008 (Figure 5.4), we observe an increasing trend that rises steeply during the post-liberalization period. Figure 5.5 presents the trend in economic inequality as reflected by the share of national income held by the top 1 per cent, the top 10 per cent, and the bottom 50 per cent of the population, organized by economic strata. Since the share of the top 1 per cent and the top 10 per cent has steadily increased, while the share of the bottom 50 per cent has continued to decrease in the post-liberalization period, we can deduce that there is rising economic inequality in the country. When juxtaposed with India's sharply rising GDP in the post-liberalization period, this diverging economic distribution pattern implies that larger slices of the fast-growing economic pie went to the upper economic strata (top 10 per cent or top 1 per cent), while for the vast majority in the lower strata (bottom 50 per cent), their share is shrinking. Against this backdrop, it is imperative to examine whether the relationship between carbon emissions and economic inequality in India changed in the post-liberalization period.

The impact of a major policy change usually becomes more evident after a time gap, and this also holds true for India's liberalization policy. We observe a steeper increase in GDP, trade, carbon emissions, and inequality in the 2000s compared to the 1990s (see Figures 5.2, 5.3, 5.4, and 5.5, respectively). Therefore, I further segmented the post-liberalization period into two decadal categories, 1992–1999 and 2000–2008, to evaluate whether the emissions and inequality relationship differed in these post-liberalization phases.

Results

Table 5.3 provides the measure of correlations between the variables used in the analysis. It shows that at the state level, carbon emissions are positively related with the Gini index of inequality, gross state domestic product, share of the industry sector, share of the service sector, population, and share of the urban population (see the 'carbon' column in Table 5.3). The scale of economic activity (state domestic product) and population size show the strongest correlation with the scale of carbon emissions. These positive correlation measures indicate that carbon emissions increased with each of these economic and demographic drivers. However, a simple correlation measure does not give us a clear understanding of the relationship between emissions and inequality. Several other factors or variables may influence the relationship between emissions and inequality, and it is important to account for those relevant drivers of carbon emissions. This can be accomplished through the method of multivariate regressions.

Table 5.4 presents the regression estimates. All the regressions use log-log specifications. Hence, the estimated coefficient represents the elasticity of carbon emissions with respect to the explanatory variable. In other words, the estimated coefficient of economic inequality measured the percentage change in carbon emissions on average when the economic inequality measure increased by 1 per cent. The regression estimates demonstrate that if we consider the carbon emissions and economic inequality relationship for the 1981–2008 period as a whole, there is no evidence of a significant relationship between the two (see column [1] in Table 5.4). In other words, the elasticity of carbon emissions with respect to economic inequality was statistically equivalent to zero for 1981–2008 as a whole.

However, a more disaggregated investigation, which compared the pre-liberalization (1981–1991) and post-liberalization (1992–2008) periods, offers insightful results. We find that the carbon emissions and economic inequality relationship was negative in the pre-liberalization period but positive in the post-liberalization period (see column [2] in Table 5.4).[5] These estimates imply that a 1 per cent increase in economic inequality was associated with an approximately 0.7 per cent decline in state-level carbon emissions on average in the pre-liberalization period (1981–1991). However,

Table 5.3 Measures of Correlation between Variables

	Carbon	Gini	GSDP	Industry	Service	Population	Urban
Carbon	1						
Gini	0.2241	1					
GSDP	0.7108	0.0917	1				
Industry	0.1183	0.2415	0.2293	1			
Service	0.2685	-0.2346	0.5482	-0.0620	1		
Population	0.7005	0.2226	0.6481	0.0294	0.1552	1	
Urban	0.1431	0.2380	0.5146	0.6139	0.3753	0.0379	1

[5] The elasticity of carbon emission with respect to the Gini index in the post-liberalization period = Coefficient of ln(Gini) + Coefficient of ln(Gini)*Post_liberalization.

Table 5.4 Regression results

	ln(*carbon*)	ln(*carbon*)	ln(*carbon*)
	(1)	(2)	(3)
ln(Gini)	0.457	-0.726*	-0.809**
	(0.1371)	(0.0705)	(0.0499)
*ln(Gini)*Post_liberalization*		1.487***	
		(0.0000)	
*ln(Gini)*Initial_liberalization*			1.024***
			(0.0000)
*ln(Gini)*Later_liberalization*			1.741***
			(0.0000)
Year	0.008		
	(0.7373)		
Post_liberalization		-4.866***	
		(0.0000)	
Initial_liberalization			-3.356***
			(0.0000)
Later_liberalization			-5.824***
			(0.0000)
F test statistics:			
H_0: coefficient of *ln(Gini)* + coefficient of *ln(Gini)*Post_liberalization* = 0		5.80** (0.0316)	
H_0: coefficient of *ln(Gini)* + coefficient of *ln(Gini)*Initial_liberalization* = 0			0.41 (0.5344)
H_0: coefficient of *ln(Gini)* + coefficient of *ln(Gini)*Later_liberalization* = 0			15.91*** (0.0015)
H_0: coefficient of *ln(Gini)*Initial_liberalization* + coefficient of *ln(Gini)*Later_liberalization* = 0			7.68** (0.0159)

Notes: *Post_liberalization* is an indicator of the post-liberalization period that takes a value 1 for the years 1992 to 2008 and 0 otherwise.

Initial_liberalization is an indicator of the post-liberalization period that takes a value 1 for the years 1992 to 1999 and 0 otherwise.

Later_liberalization is an indicator of the post-liberalization period that takes a value 1 for the years 2000 to 2008 and 0 otherwise.

Sample size: 392; Number of states: 14; *** $p<0.01$, ** $p<0.05$, * $p<0.1$.

Fixed effects estimators are reported; p-values based on robust standard errors with state-level clustering reported in parentheses.

All the regression models control for *ln(GSDP)*, *ln(GSDP)^2*, *ln(Industry)*, *ln(Service)*, *ln(Population)*, and *ln(Urban)*. The estimates are not presented here in the interest of space.

in the post-liberalization period (1991–2008), a 1 per cent increase in economic inequality was associated with an approximately 0.8 per cent increase in state-level carbon emissions on average. It is thus evident that the aggregate analysis, without the classification of the pre- and post-liberalization timeframes, masked a vital qualitative change in the emissions and inequality relationship after 1991.

Further classification of the post-liberalization period into initial liberalization (1992–1999) and later liberalization (2000–2008) phases demonstrates that the positive carbon emissions and economic inequality relationship was statistically insignificant in the initial liberalization (1992–1999) period and gained significant strength in the later liberalization (2000–0808) period (see column [3] in Table 5.4).[6] These estimates imply that while the elasticity of carbon emissions with respect to economic inequality turned from negative in the pre-liberalization period to positive in the post-liberalization period, it was statistically equivalent to zero in the initial post-liberalization (1992–1999) period. However, the elasticity increased significantly to 0.9 in the later post-liberalization (2000–2008) period, that is, a 1 per cent increase in economic inequality was associated with an approximately 0.9 per cent increase in state-level carbon emissions during the 2000s. This intensification of the positive relationship between emissions and economic inequality during the 2000–2008 period qualitatively aligns with the pronounced acceleration in India's GDP and international trade in the 2000s compared to the 1990s as a result of the increase in momentum of economic liberalization.

Discussion

My finding that the relationship between carbon emissions and economic inequality in India changed to a positive one in the post-liberalization period, and gained significant strength in the 2000s compared to the 1990s, raises the following important questions. First, what is the underlying reason for this positive relationship in the post-liberalization period? Second, what does this positive relationship imply for climate and socio-economic justice in India? This section suggests and discusses plausible answers to these questions.

---

[6] The elasticity of carbon emission with respect to the Gini index in the initial post-liberalization period = Coefficient of $ln(Gini)$ + Coefficient of $ln(Gini)*Initial\_liberalization$, and in the later post-liberalization period = Coefficient of $ln(Gini)$ + Coefficient of $ln(Gini)* Later\_liberalization$.

The positive relationship between emissions and economic inequality needs to be analysed vis-à-vis differences in access to resources and consumption patterns across economic strata. If the carbon footprint of the higher economic strata is much larger than that of the lower economic strata, then an increase in economic inequality, or greater concentration of resources in the upper strata, is expected to increase total carbon emissions. However, if the carbon footprint of the higher economic strata is much smaller than that of the lower economic strata due to the use of more efficient technologies, then an increase in economic inequality could be associated with a decrease in emissions. Hence, an increase in economic inequality may increase or decrease the scale of emissions depending upon differences in consumption patterns and the resulting carbon footprints across economic strata.

Compared to the pre-liberalization period (before 1991), when access to global markets was rather limited for all economic strata in India, the post-liberalization period represents a marked departure, as it opened up access to global products and technology, especially for the upper economic strata. Therefore, the carbon footprint of the upper economic strata increased substantively due to their enhanced access to more carbon-intense global products and technologies in the post-liberalization period. As a result, an increase in economic inequality – or a higher concentration of resources in the upper economic strata – was linked to an increase in the total state-level carbon emissions in the post-liberalization period due to a spike in the carbon footprint of the upper economic strata.

The empirical analysis I have discussed here was based on aggregate state-level data. Therefore, evaluating the differences in the carbon footprints of different economic strata within a state was not feasible. However, evidence from the existing literature demonstrates that the upper economic strata contributed more to carbon emissions in the post-liberalization period, and the upper economic strata's propensity to emit increased substantively relative to the lower economic strata in post-liberalization India. For example, Mukhopadhyay (2008) examined household-level data for the years 1983–1984, 1989–1990, 1993–1994, and 1999–2000 and found that carbon emissions accelerated in the 1990s, and the highest income groups were the prime driver of the increased emissions. Parikh et al. (2009) analysed data from 2003–2004 and found that the urban top 10 per cent income group emitted about 24 times more carbon than the rural bottom 10 per cent. Grunewald et al. (2012) studied data for the years 2004–2005 and 2009–2010 and found that the demand for carbon-intensive goods and services increased disproportionately as household affluence increased. Hence, the positive relationship between emissions and economic inequality during the post-liberalization period can be attributed to the increased carbon footprints of the upper economic strata in India.

Turning to the second question about the implications for climate and socio-economic justice, my finding that the association between economic inequality and carbon emissions turned positive in the post-liberalization period suggests that economic inequality is not only a sustainability concern from the socio-economic perspective but also a challenge for climate justice in India. Since the negative effects of rising carbon emissions are globally projected to have a disproportionate impact on the poor, this finding suggests that rising inequality and carbon emissions may reinforce and exacerbate both these problems while further undermining the well-being of the lower economic strata. In other words, the results suggest that in a business-as-usual world, socio-economic and climate justice will be even harder to realize for the lower economic strata in India. In essence, the positive relationship between carbon emissions and economic inequality in post-liberalization India implies that if left unaddressed, rising carbon emissions and economic inequality can spiral out of control, thereby threatening both environmental and socio-economic sustainability.

However, the positive relationship between carbon emissions and economic inequality also implies that inclusive economic development policies that reduce economic inequality will help mitigate carbon emissions as well. Therefore, a holistic approach towards economic development that leverages potential synergies between environmental and economic distribution policies and collective societal actions can mitigate carbon emissions and economic inequality jointly and foster climate and socio-economic justice. National and state-level policies targeting carbon emissions and economic inequality are often influenced by the upper economic strata, which shapes institutions and policies that govern natural resource use and allocation (see, for example, Boyce 1994). Hence, broad-based socio-political engagements and a willingness to approach these challenges holistically is the key to moving forward sustainably.

Conclusion

Since economic inequality has wide-ranging adverse effects, like distortions in economic development, lower contribution to public goods, erosion of trust, worse health outcomes, worse education outcomes, increased crime, and increased political instability, and the adverse impacts of carbon emission-driven climate change include loss of infrastructure, increased health risks, loss of livelihoods, food insecurity, migration, and violent conflicts, it is important to mitigate both carbon emissions and economic inequality. The fact that both carbon emissions and economic inequality have been increasing in India, and my finding that the relationship between them turned positive in the post-liberalization period, gives rise to serious concerns. Yet, the positive relationship between carbon emissions and economic inequality implies that instead of emissions and economic inequality mitigation being seen as conflicting goals, there is a potential to mitigate both jointly.

There is growing evidence that the lower economic strata contributes less to environmental degradation than the upper economic strata, and yet it is the lower economic strata that bears a disproportionate share of the impacts of environmental degradation, which begs for the need to reduce the injustice towards the lower economic strata. For example, Bhattacharya and Innes (2013) show that it is the higher economic strata in rural India that degrades vegetation but benefits more from vegetative resources. Yet, there exists a prevalent view in academic and policy discussions – the 'poverty–environment nexus' – that the poor, given their limited resources and inability to adopt environment-friendly technologies, drive environmental degradation. It also assumes that due to their heavy reliance on natural resources for survival, they get poorer when the environment degrades, thereby triggering a vicious downward cycle. It is important to recognize the fallacies in such assumptions in the context of carbon emissions as well so that climate justice is prioritized in policy formulation.

The finding that economic inequality became a key driver of rising carbon emissions in post-liberalization India highlights that the predominant policy focus on aggregate measures of economic health, like GDP, without taking into account the implications of patterns of economic resource distribution, is an inadequate approach to address critical challenges in climate and socio-economic justice. To facilitate the development of a more holistic approach towards sustainable development that fosters climate and socio-economic justice, further research needs to analyse the various pathways through which carbon emissions and economic inequality can be jointly mitigated and which pathways are more efficient.

References

Ankel, S. 2020. 'The 15 Top Countries for Billionaires, Ranked by How Many Live There'. *Business Insider*, 23 February. https://www.businessinsider.com/these-are-the-15-countries-with-the-most-billionaires-ranked-2020-2#4-india-119-billionaires-12 (accessed 19 April 2021).

Bhattacharya, H. 2020. 'Environmental and Socio-economic Sustainability in India: Evidence from CO_2 Emission and Economic Inequality Relationship'. *Journal of Environmental Economics and Policy* 9 (1): 57–76.

Bhattacharya, H., and R. Innes. 2013. 'Income and the Environment in Rural India: Is There a Poverty Trap?' *American Journal of Agricultural Economics* 95 (1): 42–69.

Boyce, James K. 1994. 'Inequality as a Cause of Environmental Degradation'. *Ecological Economics* 11 (3): 169–178.

Crippa, M., D. Guizzardi, M. Muntean, E. Schaaf, E. Solazzo, F. Monforti-Ferrario, J. Olivier, and E. Vignati2020. *Fossil CO2 Emissions of All World Countries: 2020 Report*. Luxembourg: Publications Office of the European Union, 2020. doi:10.2760/56420, JRC121460.

Das, S., G. Sinha, and T. K. Mitra. 2014. 'Economic Growth and Income Inequality: Examining the Links in Indian Economy'. *Journal of Quantitative Economics* 12 (1): 86–95.

PTI. 2020. 'Wealth of India's Richest 1% More Than 4-Times of Total for 70% Poorest: Oxfam'. *Economic Times*, 20 January. https://economictimes.indiatimes.com/news/economy/indicators/wealth-of-indias-richest-1-more-than-4-times-of-total-for-70-poorest-oxfam/articleshow/73416122.cms?from=mdr (accessed 19 April 2021).

Garg, A. and P. R. Shukla. 2002. *Emission Inventory of India*. New Delhi: Tata McGraw-Hill Publishing Company Limited.

Ghoshal, T. and R. Bhattacharya. 2008. 'State Level Carbon Dioxide Emissions of India: 1980–2000'. *Arthaniti – Journal of Economic Theory and Practice* 7 (1–2): 41–73.

———. 2012. 'Carbon Dioxide Emissions of Indian States: An Update'. SSRN paper no. 2166900.

Gill, P. 2021. 'Inequality Is Only Increasing in India — The Resulting Lack of Trust May Make Climate Change a Tougher Challenge to Tackle'. *Business Insider*, 24 March. https://www.businessinsider.in/science/environment/news/inequality-is-only-increasing-in-india-the-resulting-lack-of-trust-may-make-climate-change-a-tougher-challenge-to-tackle/articleshow/81666431.cms (accessed 19 April 2021).

Goswami, A. 2020. 'Richest 1% Emit Twice As Much Carbon As Poorest 50%: Oxfam Report'. *Down to Earth*, 1 October. https://www.downtoearth.org.in/news/climate-change/richest-1-emit-twice-as-much-carbon-as-poorest-50-oxfam-report-73612 (accessed 19 April 2021).

Grunewald, N., M. Harteisen, J. Lay, J. Minx, and S. Renner. 2012. The Carbon Footprint of Indian Households'. Conference paper for the 32nd General Conference of the International Association for Research in Income and Wealth, Boston, 5–11 August.

Grunewald, N., S. Klasen, I. Martínez-Zarzoso, and C. Muris. 2017. 'The Trade-off Between Income Inequality and Carbon Dioxide Emissions'. *Ecological Economics* 142: 249–256. http://dx.doi.org/10.1016/j.ecolecon.2017.06.034.

Intergovernmental Panel on Climate Change. 2014. *Climate Change 2014 – Impacts, Adaptation and Vulnerability: Regional Aspects*. New York: Cambridge University Press.

IQAir. 2020. 'World's Most Polluted Cities 2020'. https://www.iqair.com/us/world-most-polluted-cities (accessed 19 April 2021).

Jorgenson, A., J. Schor, and X. Huang. 2017. 'Income Inequality and Carbon Emissions in the United States: A State-level Analysis, 1997–2012'. *Ecological Economics* 134: 40–48. https://doi.org/10.1016/j.ecolecon.2016.12.016.

Kotwal, A., B. Ramaswami, and W. Wadhwa. 2011. 'Economic Liberalization and Indian Economic Growth: What's the Evidence?' *Journal of Economic Literature* 49 (4): 1152–1199.

Matthews, H. D. 2016. 'Quantifying Historical Carbon and Climate Debts among Nations'. *Nature Climate Change* 6 (1): 60–64.

Mukhopadhyay, K. 2008. 'Air Pollution and Income Distribution in India'. *Asia-Pacific Development Journal* 15 (1): 35–64.

Oxfam. 2020. 'Carbon Emissions of Richest 1 Percent More Than Double the Emissions of the Poorest Half of Humanity'. 21 September. https://www.oxfam.org/en/press-releases/carbon-emissions-richest-1-percent-more-double-emissions-poorest-half-humanity (accessed 19 April 2021).

Parikh, J., M. Panda, A. Ganesh-Kumar, and V. Singh. 2009. 'CO_2 Emissions Structure of Indian Economy'. *Energy* 34 (8): 1024–1031.

Ritchie, H. 2019. 'Who Has Contributed Most to Global CO_2 Emissions?' Our World in Data, 1 October. https://ourworldindata.org/contributed-most-global-co2 (accessed 10 May 2021).

Roser, M. and E. Ortiz-Ospina, E. (2019). 'Global Extreme Poverty'. Our World in Data. https://ourworldindata.org/extreme-poverty (accessed 19 April 2021).

The Wire. 2018. 'Nearly 81% of the Employed in India Are in the Informal Sector: ILO'. https://thewire.in/labour/nearly-81-of-the-employed-in-india-are-in-the-informal-sector-ilo (accessed 10 May 2021).

World Bank. 2021. 'GDP (Current US$)'. https://data.worldbank.org/indicator/NY.GDP.MKTP.CD?most_recent_value_desc=true (accessed 19 April 2021).

Zhang, C. and W. Zhao. 2014. Panel Estimation for Income Inequality and CO_2 Emissions: A Regional Analysis in China'. *Applied Energy* 136: 382–392. https://doi.org/10.1016/j.apenergy.2014.09.048.

CHAPTER 6

Climate Action Plans and Justice in India

Arpitha Kodiveri and Rishiraj Sen

Introduction

'Climate change seems to be the last of the priorities of the state and central government. Despite various climate plans, we continue to privatize coal and divert forest land. How does one reconcile these decisions with the objectives of the climate action plan?' asked a senior administrative officer in the Odisha Revenue and Disaster Management Department when questioned about the auctioning of new coal blocks and the state's climate action plan.[1] His grim observation points to the political and economic barriers against implementing an effective climate policy that addresses climate justice in India.

In this chapter, we argue that India's climate policy fails to adequately address difficult political questions related to climate justice and rising inequality. As our analysis of state and national climate action plans show, India's engagement with questions of climate justice remains merely symbolic. This directly follows from the country's stance in international climate negotiations, during which it has shied away from undertaking rigorous domestic climate action citing high levels of poverty and a need to focus on economic growth (Kashwan and Mudaliar 2021).

Our analysis of India's national and state climate action plans offers insights into the often-unstated normative principles that guide decision-making on climate change within the country. In this study, we demonstrate how, if at all, these action plans incorporate questions of justice and equality. We argue that most of India's

---

[1] Interview with the senior bureaucrat by Arpitha Kodiveri in August 2019.

climate action plans demonstrate a superficial understanding of socio-economic inequalities and hence fail to adequately address the disproportionate impact of climate events on the poor and marginalized.

We begin by discussing the principles that guide climate policy internationally and domestically. We then provide a critical overview of national and state climate action plans. We then scrutinize these action plans in terms of substantive equality and climate justice criteria, namely caste, gender, poverty, and co-benefits for development. We then analyse the action plans with regard to their treatment of these substantive criteria, the limitations in their approach, and possible strategies to address these limitations.

Background

Internationally, India is known to have pioneered the approach of common but differentiated responsibilities (CBDR), which allows developing countries to prioritize poverty alleviation and economic growth over climate mitigation. CBDR assigns developed countries greater responsibility in combatting climate change due to their historical emissions. This approach is justified; however, India has failed to pay the same attention to climate equity within the country (Buda 2016). As Prakash Kashwan and Parineeta Mudaliar argue:

> India has been right to raise the question of climate injustice between North and South, but climate justice within countries is equally compelling … Reversing historically entrenched socioeconomic inequalities is closely intertwined with [domestic] climate action. (Kashwan and Mudaliar 2021)

While advocating for greater responsibility of wealthier nations in the international area, India has failed to mitigate the per-capita income of the super-rich back at home. The push for CBDR internationally allows India and other developing nations to realize their energy transition faster through technology transfer and adaptation funding from the developed world. While the demand for funding from the developed world is legitimate, it needs to be accompanied with aggressive domestic efforts to reduce rising inequality (Hurrell and Sengupta 2012).

A 2007 report by Greenpeace highlighted India's failure to address climate injustices domestically (Ananthapadmanabhan, Srinivas, and Gopal 2007). The emissions of India's richest escape notice due to the low per capita emissions of India's large poor population. The CBDR principle is intended to support India's efforts to address poverty domestically; however, socio-economic and political inequalities within the country act as a barrier against achieving such outcomes. As Haimanti Bhattacharya shows in this volume, after 1991, rising economic inequality in India can be linked to an increase in carbon emissions. Addressing inequality domestically is at the heart of addressing climate change in India and should serve as the bedrock for designing climate change policy.

India's National Action Plan on Climate Change (NAPCC) is based on the co-benefits approach or the notion that climate mitigation and adaptation interventions produce development co-benefits. For example, solar energy projects help reduce energy emissions but also produce the co-benefit of increased energy security. This co-benefits approach is inherently attractive to the bureaucracy at the national and sub-national levels due to its linkage to economic growth. However, it is being used as an excuse to continue business as usual (Dubash et al. 2013).

The co-benefits approach is a form of legacy framing in that it prioritizes economic growth as an antidote to poverty. Climate change policies based on this framing presume that there is a trade-off between climate action and development and try to minimize this trade-off by identifying development co-benefits. In this sense, the co-benefits approach confounds economic development with distributional questions of addressing rising inequality, as opposed to taking meaningful climate action that simultaneously addresses socio-economic inequalities. The Indian government is now considering an ambitious net zero target following international pressure from countries in the Global North (Panwar 2021). This essentially means that India's greenhouse gas emissions will be compensated for by negative emissions, through the creation of carbon sinks.

As climate policy in India is governed by the co-benefits approach, it is useful to reflect on its relationship with existing environmental law. Upendra Baxi, in his important work on law and poverty, argues that law can be a site of emancipation and empowerment while simultaneously being a site of exclusion and impoverishment. Environmental law in India is rooted in the impoverishment and exclusion of the poor – forest-dwelling communities are deprived of their rights due to exclusionary conservation while citizens are excluded from environmental decision-making which is concentrated in the hands of the Indian state (Baxi 1979). A robust grassroots environmental justice movement has led to changes in the enviro-legal landscape, which now includes considerations of the rights of the poor. However, these legal gains are being diluted to create an enabling environment for business (Kodiveri 2016). An example of this is the proposed amendment to the Environment Impact Assessment Notification (EIA) of 2006 in 2010. EIA 2006 requires that a public hearing be held to note the opinions of those impacted by development projects prior to the granting of an environmental clearance. This provision was already not being adequately implemented, and the proposed amendments further undermined these legal gains (Bakshi 2020).[2] The design and implementation of socially just climate policies and programmes, therefore, depend quite significantly on the extent to which different groups, actors, and agencies are represented in the policymaking process.

The debate of whether climate change is better addressed through law or policy is important, but perhaps what is equally important is the need to enforce existing environmental laws. The Air Act, 1981, Water Act, 1974, Environment Protection Act, 1986, and Forest Rights Act, 2006 provide a framework to check emissions, regulate pollution, prevent deforestation, and recognize the role of forest-dwelling communities in conservation. A recent report by Chandra Bhushan and Tarun Gopalakrishnan identified key legislations, namely the Air Act, 1981, Water Act, 1974, and Forest Rights Act, 2006, that address different aspects of climate change. The report concludes that none of these laws currently contributes to ambitious climate action (Bhushan and Gopalakrishnan 2021). Addressing climate vulnerabilities and climate injustice requires synergistic coordination between environmental law and climate policy. An example of this can be seen in the relationship between the Forest Rights Act, 2006 (FRA), and the Green India Mission, which is meant to promote afforestation to create carbon sinks. These afforestation efforts often marginalize forest-dependent people who are forced out of lands that they have historically used and called their home. Further, these programmes often violate the requirement of securing the consent of the *gram sabha* or village assembly as per the FRA. This is one example of how climate action must comply with protective legal frameworks that secure the rights of the poor and impoverished (Arasu 2020).

Climate action in India: a critical overview

Internationally, India is a signatory to the Paris Agreement and has adopted mitigation and adaptation measures as per its nationally determined contributions (NDCs) (Government of India 2015). India's NDCs focus on three quantifiable goals: first, reduce the emission intensity of the gross domestic product (GDP) by 33 per cent to 35 per cent (relative to 2005 figures) by 2030; second, increase the share of renewable energy in India's energy mix to 40 per cent by 2030; and finally,

---

[2] Under the 2006 EIA notification, six major project types were exempted from holding public hearing. These included the building of area development projects and townships, projects of strategic importance, and expanding roads and highways that do not involve the further acquisition of land.

create additional carbon sinks by expanding forests and tree cover amounting to 2.5–3 billion tonnes of carbon dioxide equivalent by 2030 (Ministry of Environment, Forest and Climate Change 2015). The NAPCC has not been updated in light of these ambitious voluntary targets adopted in the NDCs.

India does not have a coherent climate change law or policy. Instead, climate action is driven by executive orders and ad-hoc documents such as climate action plans. Not much has been mentioned about the process that went into the formulation of the NAPCC, though some scholars note that it was drafted without adequate public consultation (Dubash and Jogesh 2014; Kashwan 2017). As *Down To Earth* reports, it was a quick response to international scrutiny and did not significantly engage the Prime Minister's Council on Climate Change (PMCCC):

> While [the] PMCCC had representation of diverse sectors on paper, the document's content was primarily shaped by a three-member group from within the council – the principal scientific advisor, former secretary to the then Union Ministry of Environment and Forests, and the director general of Delhi-based non-profit The Energy and Resources Institute (TERI). The final draft was prepared by the Prime Minister's Office, further limiting the significance of inputs from the council. (Rattani et al. 2018)

The international pressure to draft the NAPCC could be one reason why the plan is focused on mitigation efforts and does not pay adequate attention to climate adaptation. The plan focuses on energy efficiency, the transition to renewable energy, and afforestation instead of measures for climate adaptation.

The NAPCC was drafted in 2008 and is coordinated by the PMCCC, an ad-hoc body meant to serve as the primary institutional node in the implementation of this action plan. This gives the executive branch enormous discretion over the planning and enforcement of climate action without parliamentary and public scrutiny. For example, the PMCCC did not consult representatives from the urban poor, women workers, fisherfolk, land rights movements, and farmers' groups. Subsequently, the council's work turned out to be a technocratic exercise instead of a serious attempt to design a climate action plan that addresses India's socio-economic realities (Kashwan 2017, 194).

The NAPCC consists of eight missions that cover a broad spectrum of areas for targeted action – such as forests, the Himalayan region, energy efficiency, water, solar, sustainable habitat, sustainable agriculture and Green India Mission – and relies on specific ministries to ensure its implementation. The ministries are required to submit their proposed plans for the implementation of their assigned mission (Dubash and Jogesh 2014). For example, the Ministry of Environment,

Forests and Climate Change is the nodal ministry for implementing the Green India Mission. Some missions recommend that states be consulted when drafting policies – for example, the National Water Mission requires that states be consulted, as water is listed in the concurrent list of the Indian Constitution and states have significant policymaking authority in this sector (Ministry of Water Resources 2009). Perhaps the most important role of the NAPCC is that it provides direction for the development of State Action Plans on Climate Change (SAPCCs).

The SAPCCs were formulated based on a common framework drafted by the Ministry of Environment, Forests and Climate Change along with the United Nations Development Program in India (Ministry of Environment, Forests and Climate Change 2010). The common framework enables states to identify region-specific vulnerabilities to climate change and align regional development priorities to the national plan. The common framework document required states to undertake three activities:

1. Identify and document the climate profile of the state, which would form a baseline assessment for developing strategies
2. Conduct an assessment of the state's vulnerability to climate change
3. Assess sector-specific emissions and develop a concrete strategy to address climate change while exploring possible sources of funding to support the implementation of the action plan. (Dubash and Jogesh 2014, 4)

While the NAPCC laid down broad guiding principles for the SAPCCs like the co-benefits approach, the common framework document goes a step further and enables states to identify vulnerabilities to climate change and accordingly devise plans. It influences the process and content of the SAPCCs to a greater extent than the NAPCC (Dubash and Jogesh 2014).

Scrutinizing climate action plans for considerations of equity and justice

Despite the many weaknesses of the NAPCC and SAPCCs, these documents represent the current thinking of the central and state governments on domestic climate action in India. Moreover, the NAPCC and SAPCCs have the potential to become conduits for the creation of new norms and expectations in specific policy fields (Lagoutte, Gammeltoft-Hansen, and Cerone 2016). It is important to study such 'norm incubation' with regards to domestic climate action. This chapter aims to investigate how the baselines, norms, and expectations embedded in these documents intersect with marginalization and experiences of injustice.

With this in mind, we analysed the contents of the NAPCC and SAPCCs to understand how and to what extent they incorporate considerations of social justice in climate planning. We specifically searched for the key terms 'co-benefits', 'poor', 'equity', 'inequality', 'women', and 'caste'. These key words were carefully chosen to understand how economic inequality, class, caste, and gender are addressed in these plans.

NAPCC

The NAPCC adopts a co-benefits approach that balances development and climate priorities to realize benefits for both. As seen in Table 6.1, carbon mitigation in buildings ensures the co-benefit of energy security. Such energy savings could improve energy access for the poor, enhance air quality, and create jobs in the renewable energy sector, among others. The co-benefits approach boosts the appeal of mitigation measures, as it has the potential to improve quality of life and the environment and reduce inequality (Dubash et al. 2013). However, this approach does not provide adequate guidance on the question of who bears the burden of mitigation and adaptation and how. The final report by the expert group on 'Low Carbon Strategies for Inclusive Growth' highlights the need for a macro-level development model that considers inclusive growth alongside low carbon strategies (Planning Commission of India 2014). While the proponents of the co-benefits approach read it through the lens of inclusivity, the question remains whether it can alter the present political economy, which is dependent on fossil fuels, or if it will deepen fissures of caste, class, and gender. In Table 6.1, we provide notable quotes from our survey of the NAPCC for the substantive criteria of co-benefits, poor as representative of poverty, equity, inequality, gender, and caste.

The NAPCC identifies the poor as being vulnerable to climate change and emphasizes the need for inclusive and sustainable development as a strategy for reducing poverty. When referring to equity, the plan reverts to referencing common but differentiated responsibility, the framework for ensuring equity in combatting climate change *globally*. The NAPCC is silent on the key terms of inequality and caste – which deal more with some criteria of domestic inequities. This shows that the plan recognizes justice and equity in the arena of global governance but lacks a concerted plan to address domestic equity on the basis of caste and class. The plan, however, does identify women as being vulnerable to the adverse impacts of climate change on multiple fronts, including access to water, healthcare, and nutrition. It goes a step further and explains how women are further marginalized by adaptation efforts and calls for programmes on adaptation to be sensitive to questions of gender.

Table 6.1 Analysis of the NAPCC on social justice considerations

Criteria	Notable Quote/s
Co-benefits (13)	'Implementing carbon mitigation options in buildings is associated with a wide range of co-benefits, including improved energy security and system reliability ... jobs and business opportunities, while the energy savings may lead to greater access to energy for the poor, leading to their improvement and wellbeing. (p. 25)
Poor (7)	'Protecting the poor and vulnerable sections of society through an inclusive and sustainable development strategy, sensitive to climate change.' (p. 2)
Equity (3)	'India looks forward to enhanced international cooperation under the UNFCCC. Overall, future international cooperation on climate change should address the following objectives: • Provide fairness and equity in the actions and measures • Uphold the principle of common but differentiated responsibilities in actions to be taken, such as concessional financial flows from the developed countries, and access to technology on affordable terms' (p. 48) 'We are convinced that the principle of equity that must underlie the global approach must allow each inhabitant of the earth an equal entitlement to the global atmospheric resource.' (p. 2)
Inequality (0)	None
Women (4)	'The impacts of climate change could prove particularly severe for women. With climate change, there would be increasing scarcity of water, reduction in yields of forest biomass, and increased risks to human health with children, women, and the elderly in a household becoming the most vulnerable. All these would add to deprivations that women already encounter and so in each of the Adaptation programmes, special attention should be paid to the aspects of gender.' (p. 14)
Caste (0)	None

Source: Author's compilation based on data from Government of India (2008, 2, 14, 25, 48).
Note: * Parentheses in the 'criteria' column indicate the number of times the term occurred.

Climate justice and the state action plans

For the analysis of SAPCCs, we chose the following states: Odisha, Chhattisgarh, Rajasthan, Assam, Bihar, and Uttarakhand. The selection of states reflects their vulnerability to various effects of climate change, along with some consideration of their geographic representation. These state action plans provide a glimpse into how states have attempted to assess their vulnerability and address inequality and livelihood concerns. We also examine the case of Kerala, which has an exemplary network of civic groups and locally elected governments that enable the relatively successful implementation of state-led initiatives that promise to promote climate justice.

Caste

As can be seen in Table 6.2, the SAPCCs propose diverse strategies to address the question of caste and identify the vulnerability of SC communities based on their livelihoods. Uttarakhand, for instance, speaks to the discrimination experienced by Dalits and women, which makes them more vulnerable to the impacts of climate change. Odisha's SAPCC speaks to the challenge of rapid urbanization and the impact it will have on SC communities. Assam's SAPCC examines the link between caste and access to clean water, but stops at identifying the problem and does not propose ways to address it like the other SAPCCs examined here. However, as will be shown in the next section there are limitations in how caste is addressed in the SAPCCs on aspects of discrimination.

Gender

The SAPCCs mention these tools of integrating with existing policy and gender budgeting, but do not provide an overarching framework for responding to gender concerns. The SAPCCs address gender in various ways. Chhattisgarh addresses the question of gender by integrating its SAPCC with its women empowerment policy. Uttarakhand seeks to incorporate the tools of gender budgeting and participation of women in energy planning. Odisha addresses gender concerns within specific sectors.

Co-benefits

The SAPCCs identify climate action–development co-benefits for several sectors, though they differ in how they approach the co-benefits principle. Odisha, for example, further divides co-benefits into resilience-related and mitigation-related, thus expanding the scope of how the co-benefits principle can be deployed. Rajasthan limits the co-benefits approach to mitigation and uses greenhouse gas inventorization to assess where mitigation is occurring. The co-benefits approach as understood in these plans, as the next section will argue, fails to address rising inequality.

Table 6.2 SAPCCs and climate justice considerations

State Action Plan	Caste	Women	Poor	Co-benefits	Inequality	SC/ST (A legally recognized category within the Indian constitution)
Assam	Identifies that Scheduled Castes (SC) living in rural areas are vulnerable to climate change without access to clean water and sanitation (p. 52).	Does not mention women.	The effects of climate change will be felt most strongly by the poor. Poverty is a major challenge for Assam, as the poverty rate is 36 per cent higher than the Indian national average. Apart from economic growth, availability and access to public health services has been a challenge (p. 55).	Does not mention co-benefits.	Does not mention inequality.	In 2011—2012, as 31.98 per cent of the state's population lived below the poverty line against an all-India average of 21.92 per cent, with majority of the population, especially the people living in interior rural areas, in areas inhabited by Scheduled Caste & Scheduled Tribe population, tea garden areas and far flung 'char' (riverine) lack facilities of safe drinking water, sanitation, etc. (p. 52).

(Contd)

(*Contd*)

State Action Plan	Caste	Women	Poor	Co-benefits	Inequality	SC/ST (A legally recognized category within the Indian constitution)
Bihar	Assesses the socio-economic vulnerability of the masses in various remote locations of the state with a particular emphasis on gender, class, caste, ethnicity, physical ability, community structure, existing decision-making processes, and other local factors (p. 81).	Recognizes the different roles that men and women play in society and the unequal power relations between them. While a large number of poor, rural women depend on climate-sensitive resources for survival and their livelihoods, they are also less likely to have the education, opportunities, authority, decision-making power, and access to resources they need to adapt to climate change. Women's vulnerability to climate change differs from men, and climate change interventions that are not gender-responsive often result in deepening the existing gender divide (p. 26).	The state emphasizes inclusive development, as articulated in its Approach Paper to the 12th Five Year Plan (FYP). By extension, the state also recognizes that since climate change can disproportionately impact the poor, women, children, and the aged and can also impact livelihoods, sectoral planning under the BAPCC needs to explicitly integrate poverty, livelihoods, and equity concerns (p. 26).	Multiple 'Sector Co-Benefit Identification Studies' will take place to identify co-benefits (p. 174). The co-benefit approach is one of the tenets of Bihar's Action Plan (page xiii). The NAPCC seeks 'to promote better understanding of sectors like climate change, adaptation, mitigation, energy efficiency, and natural resource conservation while pursuing economic development resulting in co-benefits for climate change' (p. 2).	Reduced intra-state inequity between the various regions of Bihar along with reduced inter-district inequality, especially in infrastructure and service provision, as these have a bearing on livelihoods and thus adaptive capacity as well (p. 26).	Ensure social equity in distribution of assets for drinking water so that the SC/ST population and other poor and weaker sections, including minority communities, are fully covered (p. 70).

(*Contd*)

State Action Plan	Caste	Women	Poor	Co-benefits	Inequality	SC/ST (A legally recognized category within the Indian constitution)
Chhattisgarh	The combined Scheduled Tribe (ST) and Schedule Caste (SC) population is 43.37 per cent, which is one of the highest among major states, and the difference among them and other social groups in terms of assets, attainment, and access to entitlements is stark. Geographical isolation and social exclusion compound the problems in mainstreaming these vulnerable communities (p. 11–12).	The 12th FYP of Chhattisgarh will aim at addressing these development challenges by reducing inequality and deprivation and fostering an accelerated and inclusive economic growth. It will work across all sectors to further promote human development that impacts and improves the lives of all – the marginalized, women, aged, minorities, and differently abled (p. 3).	Does not mention poor.	Does not mention co-benefits.	States that in a functioning democracy, responsive governance is about ensuring symmetry of power in the elected representative-functionary-community praxis, citizen-centric administration, accountability and transparency of processes/ procedures, strong outcome orientation, and above all, delivering public goods and services in a manner that reduces inequality and vulnerabilities (p. 30).	Presently, in the 2010–2011 plan, the outlay has been increased to INR 2,985.40 lakh, and with more availability of funds due to changes of some heads from plan to non-plan, the department is planning to distribute 18,000 backyard poultry units, 500 male pigs and 500 pig trios, and 6,000 male bucks to different SC and ST beneficiaries (p. 42).
Odisha	The rural poor in Odisha depend mostly on agriculture and forest resources (especially the SCs and STs). The high level of poverty in Odisha is closely tied to the state's low productivity in agriculture (p 4). "The population of scheduled castes and tribes are higher, the inequity increases because of rapid urbanisation." (p. 28)	Therefore, gender mainstreaming requires a contextual analysis of the needs, priorities, roles, and experiences of women and men as well as the integration of specific actions proposed under the SAPCC to address any gender-related inequalities (p. 160).	Odisha is India's eighth-largest state, comprising 4.7 per cent of India's land mass, 3.37 per cent of its population (around 42 million people), and over 5 per cent of its poor (p. xi).	Co-benefits are divided into two categories: (i) resilience-related and (ii) mitigation-related in the agricultural sector (p. 58). It also discusses the co-benefits of coastal zone management (p. 65).	Does not mention inequality.	Poverty among the scheduled tribe (ST) and scheduled caste (SC) communities has been falling at a faster rate; however, the ST communities remain poorer than other social classes (Odisha Economic Survey Report, 2013–2014, 36).

State Action Plan	Caste	Women	Poor	Co-benefits	Inequality	SC/ST (A legally recognized category within the Indian constitution)
Rajasthan	Does not mention caste.	High infant mortality rate, maternal mortality rate, malnutrition among children and women, high incidence of childhood diseases, child marriage, declining sex ratio of girls under six years, low female literacy in comparison to the national average, inadequacies in water supply and sanitation, and poor health and poor socioeconomic status of women along with social discrimination are causes of concern for population health in the state (p. 94).	The state aims to implement an inclusive and sustainable development strategy that protects the poor and vulnerable sections of society from adverse effects of climate change (p. 3).	The Rajasthan Action Plan on Climate Change has been prepared 'building on the adaptation priorities and mitigation co-benefits in the state...' (p. 7). 'The Rajasthan Pollution Control Board will be guided by GHG inventorization process to plan measures and generate other co-benefits' (p. 35).	Does not mention inequality.	Does not mention SC/ST significantly.

(Contd)

State Action Plan	Caste	Women	Poor	Co-benefits	Inequality	SC/ST (A legally recognized category within the Indian constitution)
Uttarakhand	Economically vulnerable groups, including SCs and STs, have a high dependence on forest resources for the collection of fodder, medicinal plants, and firewood. These ecosystem services are highly climate-sensitive, and the regional economy is thus more vulnerable (p. 23).	Climate change will have differentiated impacts, which will be more severe for women, children, and marginalized groups in hill communities. As such, livelihood activities in the Indian Himalayan Region have a higher level of sensitivity and a disproportionate exposure to climate change. (p. 23)	Impacts of climate change are also likely to be iniquitous – the poor, women, the aged, and the very young, especially in underdeveloped or developing area contexts, are relatively more vulnerable due to their greater dependence on climate-sensitive sectors such as agriculture, fisheries, and forestry for their livelihoods or their limited adaptive capacity (p. 21).	Uttarakhand, since its creation in 2000, has set up institutions and promoted programmes that are bound to facilitate mainstreaming various adaptation measures, build the resilience of vulnerable communities and households, deepen the impacts of national missions, and provide co-benefits through developmental interventions (p. 25).	Data from the United Nations Development Programme suggest that Uttarakhand's Human Development Index (HDI) and the corresponding Inequality Adjusted Index (IHDI) stand at 0.515 and 0.345, respectively in 2013. The state's HDI and IHDI ranks 7 and 10, respectively, among Indian states. Considering the state's HDI ranking of 18 in 2005, it has made significant progress in human development (p. 35).	Mentions how the SC and ST communities livelihoods depend on ecosystem services that are climate-sensitive (p. 23)

Source: Authors' compilation based on state climate plans: Department of Environment and Forest, Government of Assam (2015, 52, 55); Government of Chhattisgarh (2013, 3–4, 11–12, 31, 42); Forest and Environment Department, Government of Odisha (2018, 4, 28, 36, 65, 160); Government of Rajasthan (2014, 3, 7, 35, 94); Government of Uttarakhand (2014, 21, 23, 25, 35).

Poverty

The SAPCCs understand poverty to be a vector of vulnerability and aim to address it through inclusive and sustainable development. The approaches mentioned here are closely aligned with the ways the SAPCCs understand inequality. Bihar identifies the poor as being vulnerable and goes a step further by incorporating sectoral planning that is sensitive to the livelihood requirements of the poor. Uttarakhand similarly deepens the understanding of the poor with a focus on the young and their dependence on climate-sensitive sectors for their livelihood. Assam's action plan highlights the issue of lack of access to good healthcare infrastructure, which renders the poor more vulnerable to the public health impacts of climate change.

Inequality

The SAPCCs vary in their understanding of inequality. In the five plans that we examined, inequality does not find mention in two of them. Bihar's state action plan focuses on inequality between districts and seeks to reduce these gaps by improving infrastructure and service delivery. Chhattisgarh's state action plan emphasizes the need for transparency and increased citizen participation in the governance process. Uttarakhand takes stock of the degree of inequality within the state by relying on the United Nations Development Programme's Human Development Indicators, including an inequality-adjusted measure of inequality.

SC/ST

The SAPCCs also take into account SC and ST communities and estimate their vulnerability while discussing the state programmes that they can access. In Assam, it speaks to the vulnerability of SC and ST communities in access to sanitation and safe drinking water. In Bihar's state action plan, what stands out is the acknowledgment of how SC and ST communities are discriminated against in accessing water and the government's aim to address it. In Chhattisgarh's state action plan, it specifically refers to these communities as beneficiaries to livestock-specific government schemes as ways of enhancing climate resilience of these communities. In Odisha's state action plan, it identifies that the rate of poverty within the SC and ST community is falling, though the STs remain poorer than other communities. In Uttarakhand's state action plan, it identifies the vulnerability of the SC and ST community based on their livelihood dependence over forest resources, which are sensitive to adverse impacts of climate change.

Analysis

As the survey of the keywords across the NAPCC and SAPCCs show, they serve as good starting points to begin thinking about climate action, but they propose limited interventions targeting climate justice. Some scholars argue that the SAPCCs serve as localized versions of climate action plans. The SAPCCs need to be considered an iterative process; the plans in their current form work as documents that lay out the broad objectives but lack a granular strategy (Dubash and Jogesh 2014). We will begin with a substantive analysis of the key terms to understand the limitations of the NAPCC and SAPCCs in this regard.

Caste and action plans

The analysis above shows a lack of serious attention to questions of caste and other forms of inequality in the NAPCC and SAPCCs. It reinforces Mukul Sharma's argument that environmentalism in India suffers from 'Dalit blindness'. Environmental movements and the discourse on environmental justice do not adequately accommodate questions of untouchability and caste-based exclusion from access to resources (Sharma 2012, 2017, 1–60). For example, Dalit communities in Kandhamal, Odisha, are dependent on access to forest produce for their livelihoods, but they are excluded from accessing these areas by Adivasi communities recently converted to Hinduism (Kodiveri 2016). Addressing discrimination against Dalits and other so-called lower caste groups in accessing resources, particularly land and water, remains an important challenge in environmental and climate justice in India (Sharma 2017, 1–60).

A study by the National Commission on Dalit Human Rights (NCDHR) showed that Dalits are vulnerable to the impacts of climate change due to loss of livelihood and lack of access to resources for climate adaptation (National Dalit Watch of National Commission on Dalit Human Rights and Society for the Promotion of Wastelands Development 2013). The SAPCCs acknowledged that SC groups, whose livelihoods depend on forest resources and agriculture, are highly vulnerable to the impacts of climate change but does not speak to the aspects of discrimination faced by these communities. It is noteworthy to see that Chhattisgarh has proposed specific schemes of agro-forestry to support the livelihood strategies of SCs. By virtue of their caste identity, Dalit communities are often denied access to resources such as land and water in India's rural and urban areas. Landlessness is highest among Dalit communities, rendering them socially and economically weaker to combat the impact of climate change on their livelihood. None of the SAPCCs speak to the need to ensure equitable distribution of land and access to water as well as commons (Thorat and Newman 2007).

Perhaps the starkest form of caste-based discrimination is experienced by Dalits who serve as sanitation workers. In Chennai, after the floods in 2018, Dalit communities were called upon to clean the entire city and get rid of the bodies. Despite providing these essential services, they were discriminated against and were denied access to food and water (Rehman 2017). Similarly, when the floods hit Cuddalore, Tamil Nadu, in 2013, Dalit communities living in low-lying areas were denied access to drinking water from neighbouring villages as the floods had damaged their homes (National Dalit Watch of National Commission on Dalit Human Rights and Society for the Promotion of Wastelands Development 2013). These examples tell us that the burdens and costs of climate change are unevenly distributed. The SAPCCs do not fully capture the complex nature of the relationship between caste-based discrimination and the impacts of climate change.

They neglect two significant aspects – the discrimination that communities considered lower in the caste hierarchy face and an intersectional understanding of the discrimination faced by Dalit women. As Behl and Kashwan argue in this volume, the intersectionality of gender, caste, and class means that poor Dalit women face the severest forms of discrimination in accessing water given increasing scarcity. This places them in a precarious situation when confronting the impacts of climate change, especially in the context of disasters. As the report by the NCDHR argues, Dalit women struggle after disasters:

> Declining food production due to climate change has turned entire populations, particularly men in the Dalit dominated village into migrants. The Dalit women are left behind and are vulnerable to greater sexual harassment. They would have to bear the double brunt of caste and gender; men are more equipped to handle situations of extreme distress as compared to women. (National Dalit Watch of National Commission on Dalit Human Rights and Society for the Promotion of Wastelands Development 2013, 26)

The SAPCCs incorporate caste as one of the relevant socio-economic parameters. However, these plans do not address caste-based discrimination, which leads to the exclusion of Dalit communities from access to basic resources. The experience of exclusion is also gendered in nature – Dalit women are more vulnerable to disasters and the livelihood impacts of climate change.

Women and the action plans

The SAPCCs identify women as being vulnerable to climate change, but the plans are not gender-responsive. The Climate and Development Knowledge Network

(CDKN), a network of organizations working to enhance the climate resilience of poor communities that will be impacted by climate change, conducted a systematic study of how gender is understood and articulated in the SAPCCs (Sogani 2016). They concluded that women are specifically vulnerable to lower food production, water scarcity, and distress migration due to climate change. The study further stated that women face a heavier burden in terms of climate adaptation because of the feminization of agriculture (also see Khadse and Srinivasan in this volume).

The CDKN's gender-responsive framework suggests that each SAPCC should collect data on the impact of climate change on women, forge strong ties with the state department of women and child welfare, harness local women's groups in tackling climate change through a bottom-up approach, and work towards standardizing gender budgeting for climate-change schemes and plans. While a gender-responsive framework provides a robust starting point, the CKDNs proposed framework views women as a homogenous group, when in reality women face different circumstances based on their class, caste, and sexual orientation. This intersectional understanding of how women experience the impacts of climate change is missing across the different state action plans (Sogani 2016).

In 2018, Kerala prepared a gender-inclusive climate action plan that identified women's vulnerability to climate change in terms of agriculture, forestry, coastal communities, water resources, disasters, and social exclusion. Kerala also addresses these vulnerabilities through its Kudumbashree Mission, which seeks to alleviate poverty by creating decentralized support networks for women. It further integrated the Kudumbashree Mission with the Mahatma Gandhi National Rural Employment Guarantee Act, 2005, as is mentioned in the gender-inclusive state action plan developed in 2018:

> The poverty eradication mission called Kudumbashree and the wage labour available under MGNREGA (employment guarantee scheme) has proved to be of help for women to get engaged in agriculture and related tasks. They have leased land and stated cultivating and a recent study has pointed out that 52,995 hectares is presently under cultivation. Most of this was land that was lying fallow. Using the employment guarantee scheme, about 300 local governments (Panchayats) have utilised the labour of women in soil conservation, recycling of plastics, and reclaiming water bodies. (State of Kerala 2018, 24)

The state government has also harnessed women's self-help groups for capacity-building for climate adaptation (Jain 2020). This initiative uses a threefold approach: recognizing the land rights of women, creating local groups of women called joint liability groups, and incentivizing organic farming to enable women to keep practising agriculture as a form of livelihood. In the district of Wayanad, the state is supporting Adivasi women by integrating the Panchakrishi programme for sustainable agriculture with the National Rural Livelihoods Mission to assist women farmers by ensuring market access and biodiversity conservation. This approach is significant in how it seeks to address the complex problem of gender-based climate vulnerability through an existing scheme (Jain 2020).

Poverty, inequality, and the action plans

K. N. Ninan argues that climate change will aggravate poverty in two ways: the population living under poverty will increase, and the conditions of those living in poverty will subsequently worsen (Ninan 2019). Haimathi Bhattacharya clearly articulates in this volume that with increased inequality, there will be a rise in emissions. This alerts us to the relationship between poverty, emissions, and climate action. Reducing inequality and poverty are thus essential ingredients of realizing equitable climate action.

Unfortunately, the political economy of India is characterized by rising poverty and inequality – India dropped a spot to occupy the 131st rank among 189 countries in the Human Development Index (United Nations Development Programme 2019). Poverty eradication programmes, particularly the National Rural Employment Guarantee Act and Food Security Act, that were meant to reduce inequality have not been adequately implemented. Similarly, as Atul Kohli argues, the Indian government is pro-business and is characterized by a narrow alliance of interests of the state and business (Kohli 2009).

The welfare state thus has been in retreat in the Indian context, given the lack of access to healthcare, education, nutrition, agricultural productivity, and jobs for large sections of the population. India has not sufficiently invested in welfare services and has chosen a path of deregulation of environment and labour laws to further the interest of big business (Jacob 2020).

As has been pointed out in the previous section, the NAPCC and SAPCCs lack an intersectional understanding of the forces and effects of the injustices and vulnerability experienced by women, Dalits, and the poor. The poor are mentioned frequently in the SAPCCs but are described as an all-encompassing and monolithic category. The state plans do not tease out the underlying conditions that push groups, individuals, and communities into poverty. A significant variation is expected in the specific ways in which these vulnerabilities manifest in different geographic, agro-ecological, and sociocultural contexts; factors of caste, class, gender, and intersectional inequalities matter everywhere. As such, any vulnerability assessment in India must account for them.

Co-benefits and the SAPCCs

Navroz Dubash and others offer a clearer path for co-benefits through their multi-criteria approach wherein they state that it must be accompanied with a clear decision-making framework that will assist states to understand the trade-offs involved, their possible impacts, and the multiplicity of factors to be considered, including growth, inclusion, and environment. They argue that low carbon growth can be achieved using a framework for decision-making called a multi-criteria analysis. This tool offers a way out of potential implementation failures (Dubash et al. 2013). In contrast, the co-benefits approach fails to challenge the political economy of extraction and rising inequality. It prioritizes economic growth as a pathway to redress poverty, while enabling the state to protect the status quo. For example, the action plans espouse renewable sources of energy for their co-benefits of cleaner air and lesser carbon emissions; however, such a selective focus on the 'benefits' of renewable energy excludes the problems of land acquisition and dispossession linked to large-scale renewable energy projects. Such a selective focus on specific benefits mitigation obscures the root causes of socio-economic and political inequalities – an extreme reliance on extractive models of development.

The SAPCCs examine how co-benefits can be achieved in sectors like agriculture, organic farming, manufacturing, afforestation, and renewable energy. These are much-needed strategies, but the action plans do not address the difficult questions of inequality and the pathway to low carbon growth. Building enduring climate resilience requires public investment in infrastructure, affordable housing, health, education, social safety nets, land redistribution, and recognition of rights to land and forest commons. These remain the most important pathways to reducing vulnerability, but the plans do not address them sufficiently.

State accountability, laws, and action plans

The plans do not offer strategies for effective enforcement of existing environmental laws. India's laws regarding air, water, and environmental protection, and those governing forests and concerning pollution and deforestation, are seldom enforced or implemented. This is a significant challenge and threat to climate change that SAPCCs do not identify. There is a need to limit the dilution of these laws and strengthen their implementation while keeping in mind the need for community participation and recognizing their rights over resources. Ensuring state accountability to these plans and laws requires citizens file public interest litigations, as the action plans do not chart out an institutional framework for monitoring and enforcement (Chatterjee 2018).

India is seeing the emergence of a nascent form of climate jurisprudence, which uses existing legislations as the basis to legally challenge state inaction. The judicial response has been uneven – at times it has pushed back against state inaction, but at others has deferred to the executive. The National Green Tribunal in Delhi has ruled in 2015 that it can be approached for violations of the NAPCC, but no cases have been filed in light of this expanded jurisdiction (National Green Tribunal, 2015). Environmental law and policy, including climate change policy, fail to address the difficult question of the rights and entitlements of the poor and equitable distribution of the burdens and costs of environmental destruction (Rajamani 2013).

An important feature of environmental governance in India has been the centralization of decision-making power and regulatory authority with the Ministry of Environment, Forests and Climate Change. This centralization is accompanied by a failure to enforce public accountability mechanisms. Thus, holding powerful political and economic actors like corporations accountable in compliance with environmental law has been difficult. India's environmental governance failures and accountability gap can be seen in the wide discretionary power and unaccountable exercise of authority by the Ministry of Environment, Forests and Climate Change. The current spate of dilution of progressive environmental laws and policies is evidence of the shrinking space for citizens to hold the state and corporations accountable (Kashwan and Kodiveri 2021).

Conclusion

In this chapter, we analysed the inclusion of justice and equity in the NAPCC and SAPCCs. The main conclusion we draw from the analysis is that they acknowledge the vulnerability of groups based on caste, gender, and poverty. However, their analyses are based on a rather superficial understanding of the production of vulnerabilities. They are also yet to offer specific strategies for addressing these vulnerabilities. Concerted action is needed to address the serious consequences of the retreat of the welfare state, which was exposed during the second wave of the COVID-19 pandemic.

The Indian government continues to pursue an aggressive development pathway marked by a dependence on fossil fuels, mining, and extractive industrial development projects that lead to deforestation, air pollution, ecological destruction, and violation of community rights. It is imperative for climate action in India to take on the difficult question of addressing the root causes of climate vulnerability, including caste-based injustices, socio-economic inequalities, and a lack of social safety nets.

In this chapter, we have shown that the national and state action plans fail to incorporate the substantive criteria of climate justice. The gaps identified are a lack of intersectionality, the need for serious treatment of inequality, and mechanisms of state accountability. Filling these gaps can offer possible avenues to inform the potential reworking of existing policy and law or shape future law and policy.

References

Ananthapadmanabhan G., K. Srinivas, and V. Gopal. 2007. *Hiding behind the Poor: A Report by Greenpeace on Climate Injustice*. Bangalore: Greenpeace India Society.

Arasu, Sibi. 2020. 'Whose Forest Is It Anyway?' *Carbon Copy*, 1 December. https://carboncopy.info/india-afforestation-schemes-udermine-forest-rights-india/ (accessed 22 April 2021).

Bakshi, Asmita. 2020. 'EIA Draft 2020: "Violation of Environmental Law Is Seen as Development"'. *Mint*, 17 August. https://www.livemint.com/mint-lounge/features/eia-draft-2020-violation-of-environmental-law-is-seen-as-development-11597593043757.html (accessed 29 April 2021).

Baxi, Upendra. 1979. 'People's Law, Development, Justice.' *Verfassung Und Recht in Übersee / Law and Politics in Africa, Asia and Latin America* 12 (2): 97–114. www.jstor.org/stable/43108795 (accessed 29 April 2021).

Bhushan, Chandra and Tarun Gopalakrishnan. 2021. *Environmental Laws and Climate Action: A Case for Enacting a Framework Climate Legislation in India*. New Delhi: International Forum for Environment, Sustainability and Technology.

Buda, Marian. 2016. 'Common but Differentiated Responsibility: International Environmental Law Principle'. *Journal of Law and Public Administration* 2, no. 4 (2016): 82–85.

Chatterjee, Trishyarakshit. 2018. 'Independent Environmental Regulation in India: Less an Authority and More a Process from Below'. *Indian Journal of Public Administration* 64 (4): 614–626.

Department of Environment and Forest, Government of Assam. 2015. 'Assam State Action Plan on Climate Change'. September. http://moef.gov.in/wp-content/uploads/2017/08/ASSAM-SAPCC.pdf (accessed 27 December 2021).

Dubash, Navroz K. and Anu Jogesh. 2014. 'From Margins to Mainstream? Climate Change Planning in India as a "Door Opener" to a Sustainable Future'. Centre for Policy Research (CPR), Climate Initiative, Research Report, CPR, New Delhi, February. https://papers.ssrn.com/sol3/papers.cfm?abstract_id=2474518 (accessed 27 December 2021).

Dubash, Navroz K., Doraiswamy Raghunandan, G. Sant, and A. Sreenivas. 2013. 'Indian Climate Change Policy: Exploring a Co-Benefits Based Approach'. *Economic and Political Weekly* 48 (22): 47–61. www.jstor.org/stable/23527912 (accessed 29 April 2021).

Forest and Environment Department, Government of Odisha. 2018. 'Odisha State Action Plan on Climate Change'. June. http://climatechangecellodisha.org/pdf/State%20Action%20Plan%20on%20Climate%20Change%202018-23.pdf (accessed 27 December 2021).

Government of Bihar. 2015. 'Bihar State Action Plan on Climate Change'. http://moef.gov.in/wp-content/uploads/2017/08/Bihar-State-Action-Plan-on-Climate-Change-2.pdf (accessed 27 December 2021).

Government of Chhattisgarh. 2013. 'Chhattisgarh State Action Plan on Climate Change'. May. http://moef.gov.in/wp-content/uploads/2017/08/Chhattisgarh.pdf (accessed 27 December 2021).

Government of India. 2008. 'National Action Plan for Climate Change'. June. http://www.nicra-icar.in/nicrarevised/images/Mission%20Documents/National-Action-Plan-on-Climate-Change.pdf (accessed 27 December 2021).

————. 2015. 'India's Intended Nationally Determined Contributions'. October. https://www4.unfccc.int/sites/ndcstaging/PublishedDocuments/India%20First/INDIA%20INDC%20TO%20UNFCCC.pdf (accessed 27 December 2021).

Government of Rajasthan. 2014. 'Rajasthan State Action Plan on Climate Change'. September. http://moef.gov.in/wp-content/uploads/2017/09/Rajasthan.pdf (accessed 27 December 2021).

Government of Uttarakhand. 2014. 'Uttarakhand Action Plan on Climate Change'. https://forest.uk.gov.in/uploads/climate_change_information/1616764235.pdf (accessed 27 December 2021).

Hurrell, Andrew and Sandeep Sengupta. 2012. 'Emerging Powers, North–South Relations and Global Climate Politics'. *International Affairs (Royal Institute of International Affairs 1944–)* 88 (3): 463–484. https://ciaotest.cc.columbia.edu/journals/riia/v88i3/f_0025424_20783.pdf (accessed 29 April 2021).

Jacob, Rahul. 2020. 'The Pandemic Will Leave India Worse with Inequality'. *Mint*, 10 December. https://www.livemint.com/opinion/columns/the-pandemic-will-leave-india-with-worse-inequality-11607531407514.html (accessed 29 April 2021).

Jain, S. 2020. 'Human Development, Gender and Capability Approach'. *Indian Journal of Human Development* 14 (2): 320–332.

Kashwan, Prakash. *Democracy in the Woods: Environmental Conservation and Social Justice in India, Tanzania, and Mexico.* New York: Oxford University Press, 2017.

Kashwan, Prakash and Arpitha Kodiveri. 2021. 'Who Will Guard the Guardians? State Accountability in India's Environmental Governance'. *Economic and Political Weekly* 56 (6).

Kashwan, Prakash and Parineeta Mudaliar. 2021. 'Resisting the Cynical Politics of Climate Negotiations'. The Planet Politics Institute, 27 April. https://www.planetpolitics.org/ppi-blog/resisting-the-cynical-politics-of-climate-negotiations (accessed 27 May 2021).

Kodiveri, Arpitha. 2016. 'Changing Terrain of Environmental Citizenship in India's Forest'. *NLS Socio-Legal Review* 12 (2): 74–104.

Kohli, Atul. 2009. *Democracy and Development in India: From Socialism to Pro-business* New Delhi: Oxford University Press.

Lagoutte, Stéphanie, Thomas Gammeltoft-Hansen, and John Cerone. 2016. *Tracing Roles of Soft Law in Human Rights.* Oxford: Oxford University Press.

Ministry of Environment, Forests and Climate Change. 2010. 'Summary Notes from the Workshop on SAPCCs'. http://moef.gov.in/wp-content/uploads/2018/01/SAPCC-workshop-summary-2010.pdf (accessed 4 May 2021).

————. 2015. 'India's Nationally Determined Contribution: Working toward Climate Justice'. October. https://www4.unfccc.int/sites/ndcstaging/PublishedDocuments/India%20First/INDIA%20INDC%20TO%20UNFCCC.pdf (accessed 20 December 2021).

Ministry of Water Resources. 2009. 'National Water Mission: Comprehensive Mission Document'. April. http://www.nicra-icar.in/nicrarevised/images/Mission%20Documents/WATER%20MISSION.pdf (accessed 27 December 2021).

National Dalit Watch of National Campaign of Dalit Human Rights and Society for the Promotion of Wastelands Development. 2013. *Impact of Climate Change on Life and Livelihood of Dalits: An Exploratory Study from Disaster Risk Reduction Lens*. https://rightsandresources. org/wp-content/exported-pdf/ncdhrclimatechange.pdf (accessed 27 December 2021).

National Green Tribunal. 2015.'Judgement in the Matter of Gaurav Bansal v. Union of India & Ors'. Original Application No. 498/2014, 23 July.

Ninan, K. N. 2019. 'Climate Change and Rural Poverty Levels'. *Economic and Political Weekly* 54 (2): 36–43.

Panwar, T. S. 2021. 'Is India in a Position to Announce a Target of Net Zero Emissions by 2050?' *Down To Earth*, 8 April 2021. https://www.downtoearth.org.in/blog/climate-change/ is-india-in-a-position-to-announce-a-target-of-net-zero-emissions-by-2050--76377 (accessed 20 December 2021).

Planning Commission of India. 2014. 'Low Carbon Strategies for Inclusive Growth'. April. https:// cstep.in/drupal/sites/default/files/2019-02/CSTEP_Low_Carbon_Strategies_for_Inclusive_ Growth_Report_2014.pdf (accessed 10 May 2021).

State of Kerala. 2018. 'Gender Inclusive State Action Plan for Climate Change'. http://www. indiaenvironmentportal.org.in/files/file/Gender%20inclusive%20State%20Action%20plan. pdf (accessed 10 May 2021).

Rajamani, Lavanya. 2013. 'Rights Based Climate Litigation in the Indian Courts: Potential, Prospects & Potential Problems'. Centre for Policy Research, Climate Initiative. Working Paper 2013/1 May. https://ssrn.com/abstract=2464927 (accessed 20 December 2021).

Rajamani, L. 2018. 'India's Approach to International Law in the Climate Change Regime'. *Indian Journal of International Law* 57 (1–2): 1–23.

Rattani, Vijeta, Shreeshan Venkatesh, Kundan Pandey, Jitendra, Ishan Kukreti, Avikal Somvanshi, Akshit Sangomla. 2018. India's National Action Plan on Climate Change Needs Desparate Repair'. *Down To Earth*, 31 October. https://www.downtoearth.org.in/news/climate-change/india-s-national-action-plan-on-climate-change-needs-desperate-repair-61884 (accessed 10 May 2021).

Rehman, Salma. 2017. 'Murky Floodwater Mixes with Casteism: Dalits Refused Relief in Cuddalore'. *Catchnews*, 14 February. http://www.catchnews.com/social-sector/chennai-s-darkest-moment-yet-dalits-in-flood-hit-cuddalore-denied-drinking-water-shelter-1449558126. html (accessed 29 April 2021).

Sharma, Mukul. 2012. 'Dalits and Indian Environmental Politics'. *Economic and Political Weekly* 47 (23): 46–52. www.jstor.org/stable/23214921 (accessed 29 April 2021).

———. 2017. *Caste and Nature: Dalits and Indian Environmental Politics*. New Delhi: Oxford University Press.

Sogani, R. 2016. 'Gender Approaches in Climate Compatible Development: Lessons from India'. Climate Development Knowledge Network, May.

Thorat, Sukhadeo, and Newman, Katherine S. 2007. 'Caste and Economic Discrimination: Causes, Consequences and Remedies'. *Economic and Political Weekly* 42 (41): 4121–4124. www.jstor.org/stable/40276545 (accessed 25 May 2021).

United Nations Development Programme. 2019. 'Human Development Index'. http://hdr.undp. org/en/content/human-development-index-hdi (accessed 25 April 2021).

स्टेज

आम्ही स्टेजवर गेलोच नाही
आणि आम्हाला बोलावलंही नाही.
बोटाच्या इशाऱ्यांनी—
आमची पायरी आम्हाला दाखवून दिली.
आम्ही तिथेच बसलो;

आम्हाला शाबासकी मिळाली.
आणि 'ते' स्टेजवर उभे राहून—
आमचे दुःख आम्हालाच सांगत राहिले.
'आमचे दुःख आमचेच राहिले
कधीच त्यांचे झाले नाही...'

—वाहरू सोनवणे

Stage

We didn't go to the stage,
nor were we called.
With a wave of the hand
we were shown our place.
There we sat
and were congratulated,
and "they", standing on the stage,
kept on telling us of our sorrows.
Our sorrows remained ours,
they never became theirs.

*—Translated by Bharat Patankar and
Gail Omvedt*

This excerpt from Waharu Sonawane's poem 'Stage' created a bit of a storm in India's activist circles. This poem and its simple, yet lyrical, translation is quite self-explanatory. Waharu is a Bhil Adivasi, poet, and long-time social activist. It is not easy to map the relationship between Waharu's poetry and activism. Seeing that Adivasis did not have leadership, even in movements that sought to speak on behalf of Adivasis, he co-founded the Adivasi Ekta Parishad (AEP). As I learned recently, in the events that AEP holds, there is a big stage, but nobody is seated on it; there is only a microphone. This reflects AEP's belief that everyone is equal, and anyone among the Adivasis can take center stage while everyone else listens attentively.

Moreover, as Waharu argued in an interview, this is a 'fight between Adivasi values and Brahmanic values—not between Adivasis as persons and Brahmins as persons. It's a fight between democracy and autocracy.' India's environmental and climate justice movements would grow stronger roots by adopting such a truly democratic approach.

CHAPTER 7

CHAPTER 7

Social Mobilizations for Climate Action and Climate Justice in India

Prakash Kashwan

Introduction

In September 2019, more than 300 representatives of farmers' organizations, trade union federations, indigenous people's organizations, fisher groups, women's organizations, environmental groups, and a few progressive political parties from Bangladesh, Nepal, Sri Lanka, and various parts of India met in Hyderabad. This four-day-long convention concluded with the founding of the South Asian People's Action on Climate Crisis (SAPACC). The delegates voiced their concerns about the anticipated effects of the impending climate crisis and 'critiqued the inadequacy of governments' policies' (Adve 2019). In the past, India's climate activists focused almost exclusively on multinational corporations and the governments of industrialized countries, who are responsible for causing the climate crisis. They argued that questioning the Indian government would 'dilute' the demand for holding industrialized countries accountable. Therefore, the SAPACC's public critiques of India and other countries in South Asia marks an important shift in the evolution of climate movements in the region.

Social movements and civil society organizations work within the complex politico-economic and institutional context of India. On the one hand, the Constitution of India is regarded as highly progressive, affording citizens a variety of civil and political rights and freedoms and a scaffolding of democratic institutions that are functional to some extent. This context is particularly conducive for the functioning of civil society institutions that focus on relatively less controversial and apolitical questions, for example, Gandhian organizations dedicated to the 'welfare' of the poor, or those promoting tree-planting programmes. On the other hand, organizations advocating for the rights and entitlement of the poor, and those demanding effective enforcement of constitutional provisions and a welfare state, often confront a state that is extremely opaque and highly vindictive (Banerjee 2008). This 'Janus-faced nature of the postcolonial state' explains why some types of environmental movements thrive in Indian society while others face violent threats (Kashwan 2017, 10). Yet these contradictory workings of the Indian state must be understood in the context of global capitalism and its domestic beneficiaries. Instead of weakening state control in the wake of economic liberalization in the early 1990s and beyond, the Indian state has transformed into a highly centralized and extractive state that abuses its authority blatantly to selectively reallocate land and other natural resources (Rajan 2011).

This chapter situates the emerging climate justice movements in India in this broader political and economic context and the long-standing patterns of state power that led us to the present moment. Primarily, it examines three streams of social mobilizations: (*a*) conventional climate activism in India, focused mainly on the Global North and large corporations, (*b*) various people's movements that have advocated for holding governments in both the Global North and South accountable for their failure to address the environmental and climate crises, and (*c*) contemporary climate movements, including the youth climate movement in India. I bring together these three strands to investigate how their confluence may reshape climate politics and the pursuit of climate justice in India. Toward this end, this chapter analyses the political implications of the various forms of environmentalisms in India (Mawdsley 2004; Baviskar 2019). It also scrutinizes the claim that climate change presents a fundamental challenge to India's environmental movements (Lele 2012; Swarnakar 2019).

A key insight presented in this chapter is that scholarship on both Indian environmentalisms and Indian climate movements requires a more nuanced and fuller engagement with politics. This includes the multiple ways in which environmental and climate movements respond to and engage with policymaking processes and the institutional structures of the state. The next section outlines the key arguments concerning the political entanglements and draws implications for environmental social movements. It develops an analytical lens to examine how movements deploy a plethora of skills, resources, and narratives in different political spaces, both nationally and internationally. The third section uses this lens to examine how three of India's best-known environmental movements deployed various strands of environmentalisms and how it affected for their key constituents.

This is followed in the fourth section by a discussion of the three strands of climate movements in India and their likely consequences for the pursuit of climate justice. The concluding section synthesizes insights from analyses of India's environmental and climate movements to reflect on challenges concerning political accountability in India.

The politics of environmentalisms in India: an analytical lens

The scholarship on environmentalisms in India has contributed varied approaches to understand social action concerning the environment. The most prominent of these concepts is 'environmentalism of the poor', which is defined as 'actions and concerns in situations where the environment is a source of livelihood' (Martinez-Alier 2014). In the face of increasing threats to the environment and natural resources, people whose livelihoods depend on these environmental resources are likely to mobilize in favour of environmental protection. This has prompted some to refer to this form of mobilization as 'livelihood environmentalism' (Ramesh 2010). The environmentalism of the poor is often juxtaposed against 'elite environmentalism', which involves 'a class of ex-hunters turned conservationists belonging mostly to the declining Indian feudal elite and ... representatives of international agencies' (Guha 1989, 3). These networks of transnational elites advocate for an elite environmentalism that is structured to 'transplant the American system of national parks onto Indian soil' (Guha 1989, 3). This model of elite environmentalism has been adopted quite fervently by India's burgeoning middle classes who seek to mimic the lifestyle of middle-class Americans and consider weekend trips to national parks in SUVs (sport utility vehicles) as an indication of their environmental commitment (Mawdsley 2004).

Amita Baviskar has further developed and broadened these arguments in her work on bourgeois environmentalism, specifically in the context of urban environmental campaigns. Baviskar defines bourgeois environmentalism as 'the (mainly) middle-class pursuit of order, hygiene and safety, and ecological conservation ...' (Baviskar 2019, 110). This form of environmentalism emphasizes a "clean and green" environment, aesthetically slick and sanitized – without looking at one's complicity in creating environmental problems in the first place' (Ganesan 2020). Here, the middle classes mobilize universalistic discourses of 'citizenship', 'civic concerns', and 'public interest', but with the very specific intent of excluding the poor (Baviskar 2019). As such, the concept of bourgeois environmentalism draws attention to the influence of the multiple layers of sociocultural reality that shape the environmentalism of India's influential middle and upper classes.

Building on this rich scholarship on various forms of environmentalism, this chapter seeks to bring a sharper focus to the political dimensions of the different types of environmentalisms. Its approach is inspired by the vast scholarship on social movements in India and abroad (Swain 1997; Ray and Katzenstein 2005). As several of the contributors to the volume edited by Ray and Katzenstein (2005) argue, the Indian state has always exercised a very strong influence on the functioning of social movements and non-governmental organizations (NGOs). The (im)balance of power between state actors and society has changed significantly from the Nehruvian era's tolerance of social movements to the blatantly authoritarian regime that is in power now (Kashwan 2014; Sud 2020).

The foundations of Indian bureaucracy were laid during colonial rule when bureaucratic structures were not intended to be accountable to society at large. As a result, India's bureaucracy is considered 'over-developed', with very little social control and democratic accountability (Haque 1997). This incongruence between bureaucratic powers and democratic control has only widened in the post-independence era, as popular access to state apparatus has become a means of social power. This also means that civil society organizations in India are less likely to be effective compared to those in countries with relatively stronger mechanisms for state accountability. Understanding the drivers for success of environmental and climate movements requires a deeper analysis of their relation with state institutions, including the judiciary and administrative apparatus responsible for upholding environmental and climate regulations. While mass social movements are no match for the unaccountable and unforgiving authority of the Indian state, some movements have scored important successes, especially via judicial interventions in cases such as the Samatha judgment, the Niyamgiri judgment, and the judgments in response to the legal advocacy pursued by Narmada Bachao Andolan (NBA) (Banerjee 2008). We need to evaluate the strategies adopted by environmental social movements, and the extent to which they have been successful, in the context of the highly asymmetric power of market and state actors.

Two analytical strategies are central to the arguments I make in this chapter about the politics of the different forms of environmentalism. The first of these relates to the Habermasian 'public sphere', which is defined as 'the social space in which different opinions are expressed, problems of general concern are discussed, and collective solutions are developed *communicatively*' (Wessler and Freudenthaler 2018, italics added for emphasis). This chapter argues that the scholarship on the different varieties of environmentalism must build on, but go beyond, an investigation of the nature of the 'public sphere' (Baviskar 2019, 110). Even in the best of circumstances, articulation of grievances in the public sphere is merely the first step. Such grievances must then be addressed through prioritization and allocation of resources, which is the domain of politics beyond the narrow debates of the public sphere (cf. Mehta 2013). Knowing this and intent on serving their constituents, social movements strategize to respond to a given political environment. Such strategies could range from increased efforts to strengthen their grassroots presence, networks, and popularity to appealing to middle-class urban supporters, who, until recently, were relatively well-protected against the oppressive tactics of the state (Sinha 2021). Such a contextualized approach to judicious decision-making takes on board social, cultural, and political factors and processes that shape the decision-making of movements and counter-movements (Koopmans 2005).

This chapter broadens existing analyses of environmental and climate movements by focusing on the extent to which movements enter and navigate various political spaces and processes. It also recognizes that social hierarchies and the socio-economic status of movement participants shape movements strategies. This allows for the possibility that a movement's outcomes can have very different meanings and implications for various groups within a diverse pool of supporters and followers. One central argument is that movements may adopt multiple frames, environmental discourses, and political strategies, some of which may seem contradictory to an external observer. The next section applies this approach to investigate three of India's most prominent movements.

India's environmental movements and environmentalisms

Three of the most renowned environmental movements in India are the Chipko movement, the Silent Valley movement in Kerala, and the Narmada Bachao Andolan (NBA) movement against the Sardar Sarovar Project (SSP). They offer a useful snapshot of the varying ways in which movements engage with different political spaces using diverse frames that resonate differently with various sections of society. To be clear, my goal is not to present an exhaustive analysis of these movements or to argue that political processes were the only determinants that shaped their outcomes. Instead, it is to demonstrate that in addition to the most commonly talked about factors, political factors also had long-lasting consequences for the movement's supporters and participants.

Chipko movement, Uttarakhand Himalaya

Chipko (literally, 'hug the trees') was spearheaded by Dasholi Gram Swarajya Mandal (DGSM), a local Gandhian organization founded in 1964 with the mission of establishing forest-based enterprises that create local employment opportunities. However, the forest policy favoured city-based contractors over grassroots groups such as DGSM. The proverbial straw that broke the camel's back was the forest department's refusal to grant DGSM the permission to harvest ten ash trees that they could use in a workshop on making the tools needed for subsistence farming. Shortly afterward, in March 1973, the forest department allowed a sporting goods company to harvest 300 ash trees from the same forest, which triggered what metamorphosed into the famous Chipko movement (Jain 1984). The main organizers were individuals involved in local production forestry and wood-processing work who aimed to secure local control of forests for forest labour co-operatives (FLCs). While women did play an important role in the movement, their central concerns were not dramatically different – they mobilized to demand local control over forest resources that are crucial for small-scale farming and forest-based subsistence. Yet Chipko is often portrayed as an environmental or ecofeminist movement (cf. Rangan 1997). Neither of these terms is an accurate depiction of the grassroots movement that the men and women of Uttarakhand Himalaya began.

Chipko's most widely known leader, Sundarlal Bahuguna, was a timber contractor who transformed himself into a radical green leader. Scholars suggest that this move was linked possibly to Bahuguna's realization that this would help him get closer to Prime Minister Indira Gandhi, whose love for the environment was well-known (Sinha, Gururani, and Greenberg 1997). Eventually, Bahuguna and Indira Gandhi's non-environmental considerations of social and political influence resulted in a 15-year ban on harvesting trees in the Uttarakhand hills, which, in turn, brought international fame to the movement (Baviskar 2005, 165–166). The frames and strategies that Chipko's male leadership employed had significant negative consequences for the men and women of Uttarakhand Himalaya who participated in the grassroots protests. This is evident from interviews that Gayatri Devi, one of the prominent women leaders and the president of the Mahila Manga Dal of Doongri village, gave many years later.[1] Responding to a question from the environmental weekly, *Down to Earth* – which asked her 'What did you get out of Chipko?' – Devi said,

> … we never got anything out of it. The road to our village is yet to be constructed and water is still a problem. Our children cannot study beyond high school unless they can afford to go and stay in a town. The girls simply cannot do that. Now they tell me that because of Chipko the road cannot be built because everything

---

[1] Gayatri Devi was in Delhi to receive Government of India's Vrikshamitra Award in 1986 for the role that villagers of Doongri played in Chipko.

has become *paryavaran* (environment) oriented nowadays. *Hamare haq haqooq cheen liye gaye hain.* (Our rights have been snatched away) … My first fight will be for the road, *paryavaran wale chahe kuchh bhi kare* (and let environmentalists do what they will). (Mitra 1993)

Two of the key local women leaders of Chipko, Gayatri Devi and Gaura Devi of Raini, have stated ambivalent views about Chipko's success (Linkenbach 2001). Their testimonies, as well as those of others, suggest that the popular narratives surrounding the Chipko movement were influenced quite significantly by the political ambitions of its male leaders. To be clear, rural women did lead local mobilizations, but the rationale and arguments of the local women leaders were not the ones that dominated headlines. One of them explicitly denied ever having hugged trees. In at least one instance, Bahuguna is alleged to have presented a random woman as one of the leaders of the Chipko movement (Linkenbach 2001). Chipko was hyped nationally and internationally because of its appeal as a purist environmental movement (Rangan 2000). Yet the narratives of grassroots environmentalism popularized by the Chipko leadership have been deployed by other environmental movements and middle-class environmental activists.

The creation of the Nanda Devi National Park and Biosphere Reserve in the heart of Chipko land has exacerbated feelings of disenchantment among the local people, especially the Bhotias (Dogra 2002). While much of the natural sciences literature on the reserve makes clichéd references to the local communities' love for the environment, the Bhotias deployed the Chipko narratives to contest the exclusionary park-based model of conservation. Moreover, they proposed a new model of community-based tourism for 'the transformation of our region into a global centre for peace, prosperity and biodiversity conservation' (Bosak and Schroeder 2004, 6). Many Bhotias also circumvent the park-related restrictions to collect and market cordyceps, a medicinal fungus that grows in high-elevation meadows in the region (Caplins, Halvorson, and Bosak 2018). For some local community groups, the pendulum of Chipko history has swung back to where it started – that is, to the assertion of local rights to and control over the region's natural resources.

Silent Valley movement

The Silent Valley movement was led by scientists, teachers, and professionals and enjoyed significant popularity among the middle and upper classes (Jasanoff 1993). The Kerala Sastra Sahitya Parishad (KSSP), the grassroots science literacy group that spearheaded the movement within Kerala, sought to arm people with the 'weapon of scientific knowledge' to help them overcome their state of underdevelopment (Menon 2012). The Silent Valley movement leveraged social and political mobilization to ensure environmental protection for the valley and contest the feasibility and desirability of new hydroelectric dams. Though restricted primarily to the middle and upper classes, the movement yielded benefits for ecology and society, including for the poor fisherfolk and peasants living on the banks of the Kuntipuzha river. Kuntipuzha is the only undammed perennial tributary of the Bharathapuzha river, with upstream catchments originating in the Silent Valley and Mukurthi National Parks (Shaji 2015). By some measures, it is reasonable to refer to Silent Valley as a 'people's movement that saved a forest', as argued by Shekar Dattatri, whose documentary on Silent Valley brought him international fame (Dattatri 2015). The then prime minister, Indira Gandhi, is also credited for the movement's success. However, Gandhi's support for Silent Valley was hard-won.

N. D. Jayal, the then joint secretary for forests and wildlife, played a crucial role. During a visit to the valley, Jayal watched a slide show on the richness of the valley's flora and fauna, which made him sympathetic to the cause. However, his realistic assessment was that the demands to save it 'would cut no ice with the government' (Warrier 2018). So, instead of working through the government machinery, Jayal requested the famed ornithologist Salim Ali to intervene. Ali wrote to Prime Minister Indira Gandhi, who was also 'overwhelmed by letters from overseas' (Ramesh 2017, 271). This included a letter from the director-general of the International Union for Conservation of Nature (IUCN), who received a prompt response from Gandhi decrying 'a Marxist Government in Kerala which is anxious to go ahead with the project' (Ramesh 2017: 270). Gandhi set up the M.G.K. Menon Committee to investigate the matter. In its final report, the Menon Committee expressed serious concerns about the proposed dam because of the threat it posed for the valley's ecological diversity. The success of the Silent Valley movement was because of the strong support it received from India's middle-class environmental activists and political elites. Such support notwithstanding, even under the greenest PM India had, it took the combined might of highly influential public figures, both within the government and outside of it, to stop the dam.

Narmada Bachao Andolan (NBA)

The NBA was founded in 1985 with the explicit goal of securing the rights and livelihoods of people affected by the Sardar Sarovar Project (SSP). In 1985, the World Bank approved $450 million for the SSP, which was the largest of dozens of large dams planned under one of the world's largest multipurpose projects, the Narmada

Valley Development Project (NVDP) (Mathew-Shah 2015). The NBA's multifaceted mobilization, as I shall discuss further, forced the World Bank to exit the SSP in 1993 and constitute a World Commission on Dams (WCD) in May 1998, with Medha Patkar as one of its members (Vombatkere 2016). The NBA also successfully mobilized the government to establish a policy framework for the rehabilitation and resettlement of project-affected people. This did not prevent the construction of the dam, which was completed in 2015 (Satheesh 2019).

The NBA started off with demands for fair resettlement and rehabilitation, but it also took up some environmental demands with the aid of influential environmentalist groups within India and abroad. Some scholars have argued that the NBA successfully brought together 'the red politics of class struggle ... and the green politics of preserving and conserving the environment' (Ganesan 2020). However, others argued that NBA's alliances with international NGOs *shifted* its focus 'from rehabilitation and resettlement to environmental sustainability ... [which] made the movement internationally visible ... (Shah et al. 2019, 20). They argue that large and resourceful environmental NGOs headquartered in the Global North 'privileged the "green" component at the cost of the "red"' (Shah et al. 2019, 20). Indeed, the NBA's decision to include an environmental agenda resonated with middle-class environmental sensibilities at home and abroad. However, the NBA's reliance on middle-class Indian activists and global environmental groups must also be seen in the context of the Indian state's refusal to engage with its demands seriously (for an extensive discussion, see Banerjee 2008). Moreover, one cannot ignore the fact that the support of national and international environmental groups was instrumental in forcing the World Bank to initiate a series of reforms that also prompted the Government of India to develop a rehabilitation and resettlement policy framework.

The increasing influence of urban and cosmopolitan activism also gave rise to questions of representation within the NBA. A prominent local Adivasi leader asked if 'to say yes to everything that is said, to participate in activities, fill water in tubs, sweep the floor, cook food, wash utensils, carry news about NBA activities to villages, wash other peoples (*sic*) clothes; are these the main task for Adivasi activists? Is this the Adivasi leadership?' (Dwivedi 1998, 176). When another Adivasi activist asked one of the top NBA leaders 'Why are there no Adivasis in the NBA leadership?' his response was: 'Our village-level leaders are all Adivasis' (Omvedt 1997). Such a response validates the complaints raised by the local Adivasi leader mentioned above. The NBA's support base was also entangled in local caste hierarchies as 'the *majority* of affected villagers who were active participants in the Andolan consisted of relatively prosperous upper-caste farmers from the fertile plains' (Baviskar

2019, 32, italics added for emphasis). Yet the movement's metropolitan supporters projected that the movement predominantly comprised

> ... hill adivasis, picturesque in their traditional clothing, holding bows and arrows, defending a lifestyle based on benign co-existence with nature. Such performances portrayed adivasis as 'ecologically noble savages' such that saving them was coterminous with saving the river and forests. (Baviskar 2019)

This validates the argument that NBA mobilization tapped into middle-class and 'Western' conceptualizations of a marginalized community fighting to protect the pristine environment within which they live. In India and elsewhere, Adivasis and indigenous people have been stereotyped and essentialized within environmental conservation, which simultaneously continues to pay homage to indigenous rights to territorial sovereignty (Kashwan 2013; Sinha, Gururani, and Greenberg 1997). Yet it is difficult to draw strong inferences in an abstract analysis such as this. We must assess the NBA's success in addressing the multiple challenges that it confronted within the context of repressive state responses and limited resources to sustain a mass movement, especially if the movement had to maintain its support among Adivasi peasants (Banerjee 2008).

Indian environmentalism: what succeeds?

The discussion above shows that three of the most celebrated environmental movements in India relied very heavily on middle-class and international supporters who prioritized environmental concerns over the subsistence interests of local communities. However, each of these movements confronted very different circumstances, which is why their comparative analysis offers important lessons.

Silent Valley was a middle-class environmental movement that did not address questions of social justice (Omvedt 1987). Most importantly, the movement had support at the highest level of the political establishment. The male leaders of the Chipko movement portrayed it as a women's tree-hugging movement, an image that was central to academics and activists presenting Chipko as an instance of ecofeminism (for an extensive critique, see Rangan 2000). And finally, some activists within the NBA deployed simplistic and essentialized images of hill tribes living in harmony with nature, but such discourses were also accompanied by strong arguments in favour of securing local communities' rights to natural resources. However, contrary to the other two movements, the NBA faced a hostile state, which used extra-legal violence against NBA supporters and allied groups (Banerjee 2008).

Paul Routledge makes a strong argument for the scholarly responsibility to appreciate the imperatives the NBA faced to present an 'unambiguous ... public image' after smoothing over 'complexities and nuances within everyday realities in the Narmada valley' (Routledge 2003, 266). Tania Murray Li makes a related argument, suggesting that many social movement strategies are similar to 'creating an ant path', allowing them to 'push boundaries, opening up the terrain for progressive politics ... while operating within the lines of intelligibility of transnational donors or government departments' (Li 2014: 229).

These questions regarding the broader context dominated by a state co-opted by crony capitalists, the specific nature of state–movement relations, resource mobilization, representation, and political strategies should be at the core of any investigation of contemporary climate movements, which will invariably face similar challenges.

Climate movements and climate justice in India

For the better part of a quarter century of global climate negotiations, every key constituency in India has presented a unified position regarding India's climate strategy. Ironically, even as civil society lent its support to climate nationalism, the Government of India's position went through a gradual but perceptible change, especially after the failed Copenhagen Conference of Parties (CoP). At the Cancun CoP, India's Minister for Environment and Climate Change, Jairam Ramesh, made an 'impromptu addition' to his address at the high-level segment of the climate talks: 'All countries must take binding commitments under appropriate legal form' (Ramesh 2015). While there was much consternation at this changed stance, none of this has led to meaningful climate action. The following discussion illustrates that the history of activism for domestic climate justice is much longer and its roots are much deeper than is sometimes apparent from the present scholarship on climate governance in India.

History of domestic climate justice activism in India

Many of India's social activists joined hands to form the Indian Climate Justice Forum (ICJF), a coalition of Indian and international groups that mobilized on the occasion of the United Nations Framework Convention on Climate Change's (UNFCC) Conference of the Parties-8 (CoP8) meeting in October–November 2002 in New Delhi. It was meant to be a platform for marginalized groups and their representatives who are often left out of United Nations (UN) negotiations. The ICJF organized a Climate Justice Summit, which was attended by the National Fishworkers' Forum from Kerala and West Bengal; farmers from the Andhra Pradesh Vyavasay Vruthidarula Union (Agricultural Workers and Marginal Farmers Union); Adivasis representing the NBA from the Narmada Valley; indigenous people of the North-East states; and representatives from disaster prone areas in Orissa. Participants at the summit highlighted inequities within India – such as the instance of migrant workers in Delhi who came to the city to work as rickshaw-pullers and construction workers because they had been displaced by coal mining, floods, and drought. These rickshaw-pullers were the target of middle-class environmentalist campaigns to bring order to Delhi's streets (Baviskar 2019). Reflecting on the ongoing controversy, a rickshaw-puller commented, 'The rich people drive around this district of Delhi one person to a car – they are contributing to the pollution. We do not make any pollution yet we are banned from … work' (Khastagir 2002).

The risks of market actors making inroads into global climate governance, which these protests in Delhi flagged, turned out to be quite prescient. The Bali Action Plan, which was agreed upon at CoP13 in December 2007, catalysed a reliance on markets, ostensibly with the goal of promoting cost-effective climate mitigation. The Bali Action Plan institutionalized the use of forests as a means of climate mitigation by '[r]educing emissions from deforestation and forest degradation (REDD) in developing countries' (UNFCCC 2007). The Bali conference also proved to be a fountain of global and transnational climate justice mobilizations in various parts of the Global South. Soon after, activists engaging with issues in various sectors of the economy founded the India Climate Justice (ICJ) collective. The ICJ drew inspiration from the deliberations of the Durban Coalition for Climate Justice. The ICJ newsletter, *Mausam* (literally, 'weather' in Hindi), published with the support of UK-based activists Jutta Kill of FERN and Larry Lohmann of Cornerhouse, UK, sought to start a public conversation on climate in India.[2] *Mausam* debuted with a sharp critique of the dominant framework of international equity as the sole measure of climate justice:

> You cannot deny a sovereign nation its developmental energy, and the necessary, absolutely necessary, emissions, argues the government. The mainstream media; the political, scientific, and economic fraternities; and many 'responsible' NGOs echo the view. Yes, there is a climate crisis. But we did not create it, and necessary adaptation and mitigation measures will be taken; a national climate action plan

---

[2] These groups included the North Eastern Society for the Preservation of Nature and Wildlife (NESPON), the National Forum of Forest People and Forest Workers, and Nagarik Mancha, Kolkata.

is on board. ... Yes, but who are 'we'? Who 'are' the nation we celebrate? What defines 'development'? (Ghosh 2008, 2)

These climate activists did not necessarily see a conflict between the goals of international and domestic climate justice or fear that demanding comprehensive global and national climate action would overshadow longstanding environmental struggles within the country (cf. Dubash 2019). For example, the articles published in the very first issue of *Mausam* criticized India's increasing energy consumption, cautioned about biofuel's detrimental effects on India's commons and commoners, and exposed the 'scam' of CDM (the Clean Development Mechanism under the Kyoto Protocol).[3] Despite levelling a strong critique of both global and national policies, these groups continued to strongly support the principle of Common but Differentiated Responsibility (CBDR). Yet the more eclectic positions these groups developed were not sufficiently represented in international forums, in part because transnational networks, like the Climate Action Network, only supported groups that aligned with discourses of international equity.[4]

Indian climate movements today

The post-Paris scenario of global climate (non)action seems to vindicate most of the arguments that climate justice groups like ICJF made over a decade back. This includes the argument that the Government of India should be held accountable for its environmentally destructive models of development. Some of the key figures in the ICJ collective are also involved in the founding of the SAPACC, which seeks to foster science-based climate action by engaging with core constituencies within key sectors in the Indian economy. The first elected coordinator of SAPACC, Sudershan Rao Sarde, is the former director of the South Asian Regional Office of the International Metalworkers Federation (IMF). The first SAPACC convention, held in September 2019, saw a significant participation of union leaders; they recounted the wise words from Sharan Burrow, general secretary of the International Trade Union Confederation (ITUC), who often says to her fellow trade unionists: 'There are no jobs on a dead planet' (Adve 2019).

Not all mainstream trade union leaders are on board, though. Some of them argue that their members are unwilling to support the SAPACC position that India needs

---

[3] Past and current issues of *Mausam* are filed at http://www.thecornerhouse.org.uk/resources/results/mausam%20taxonomy%3A14 (accessed 26 December 2021).

[4] Personal video interview, India Climate Justice activist, 10 June 2020.

to transition from coal and other fossil fuel–based energy sources to renewable energy.[5] The positions of various workers across India are very different from those of organized trade unions, who continue to stand by longstanding notions regarding India's supposed national interests. However, the trade union sector has also expanded, with the emergence of new unions such as the New Trade Union Initiative (NTUI) founded in 2002. The NTUI seeks to provide an independent, democratic, and militant voice to working people in India. It is a member of the Trade Unions for Energy Democracy (TUED).[6] The NTUI links emission reductions with questions of social justice and development. It advocates for better regulation of the energy sector to address the climate crisis, while also pursuing 'a transition that recognises the development needs of people in the South and the key role of Labour in this process' (Mathews, Barria, and Roy 2016).

The recent uptick in youth movements in response to the climate crisis seems to be a promising avenue for a new wave of mobilization. While social media offers a low-cost and user-friendly way of making an immediate connection with the youth, India's gaping digital divide means that a heavy reliance on social media is likely to produce a movement that is skewed toward upper-middle- and upper-class youth. These biases manifested during a march to protest the government's failure to act in response to the ongoing climate crisis and stem environmental degradation. A man employed as a security guard asked a young activist carrying a placard and shouting slogans as part of the march, '*Yeh morcha kis liye kar rahe ho aap log*? (What are you guys marching for?).' The activist struggled to articulate the core message (Joshi 2019). While Joshi attributed this awkward situation to a 'language gap', the gap between India's young climate warriors and the majority of India's population is more substantive.

Notwithstanding the aforementioned gap, India's youth environmental and climate movements are already making an impact. Youth movements spearheaded social mobilization contesting the Indian government's efforts to dilute Environmental Impact Assessment (EIA) guidelines. In response, the union environment minister, Prakash Javadekar, complained about receiving 'multiple emails with the subject name similar to "EIA 2020"' (Agarwal 2020). The Delhi

---

5 Personal video interview, India Climate Justice activist, 10 June 2020.

6 TUED is a global, multi-sector initiative working to advance democratic direction and the control of energy in a way that promotes solutions to the climate crisis, energy poverty, the degradation of both land and people. It responds to the attacks on workers' rights and protections. http://unionsforenergydemocracy.org/about/about-the-initiative/ (accessed 26 December 2021).

police blocked the websites of the Let India Breathe campaign for 26 days and of Fridays for Future India (FFFI) for two weeks, apparently because they found their contents 'objectionable', depicting an 'unlawful or terrorist act', and proving 'dangerous for the peace, tranquility and sovereignty of the [*sic*] India'. Moreover, the service providers for these websites were issued a notice under the draconian Unlawful Activities (Prevention) Act (UAPA) (Adve 2021). The Indian government's repressive tendencies came to the fore recently when Disha Ravi, a Bangalore-based activist of the FFFI, was arrested for editing a Google document circulated by youth climate activist Greta Thunberg for mobilizing support for farmers protesting against the Modi government (*The Wire* 2021). Disha Ravi's arrest had a 'chilling effect' on youth movements (Rakesh 2021). However, Ravi has led from the front, releasing a brave and insightful statement after she was granted bail in the Thunberg dossier case. It is worth quoting the following long excerpt from this statement:

> I also realized, during my time in custody, that most people knew little or nothing about climate activism or climate justice. My grandparents, who are farmers, indirectly birthed my climate activism. I had to bear witness to how the water crisis affected them, but my work was reduced to tree plantation drives and clean-ups which are important but not the same as struggling for survival. Climate Justice is about intersectional equity … It is a fight alongside those who are displaced; whose rivers have been poisoned; whose lands were stolen; who watch their houses get washed away every other season; and those who fight tirelessly for what are basic human rights. We fight alongside those actively silenced by the masses and portrayed as 'voiceless,' because it is easier for savarnas to call them voiceless. We take the easy way out and fund saviourism rather than amplify the voices on ground. (*News Minute* 2021)

India's youth movements, especially under the leadership of young activists from diverse backgrounds, is a source of hope for India's environmental and climate movements.

Conclusion: toward transnational mobilizations for accountability in climate action

The history and evolution of India's environmental and climate movements, which I have surveyed in this chapter, demonstrates one undeniable fact. Middle-class or bourgeois narratives of environmentalism are extremely popular, both at home and abroad, especially among those who hold power in the status quo. However, as my analysis shows, in each instance, it also produces negative consequences for marginalized groups whom the movements seek to serve. It also shows that middle-class environmentalism is not just an urban phenomenon. On the other hand, elements of middle-class environmentalism in rural and forest contexts are entangled in 'new traditionalist discourses' that essentialize 'local communities' to appeal to romantic notions of rebellious and virtuous peasants standing up to exploitative market and state actors (Sinha, Gururani, and Greenberg 1997). Despite these and other longstanding critiques of 'the local trap', in which outsiders falsely assume that 'localized decision-making is inherently more socially just or ecologically sustainable', the romantic and essentialized portrayal of 'community' has endured within Indian environmentalism (Purcell and Brown 2005).

These discursive frames led to outcomes that reinforced inequalities within the complex field of transnational advocacy. For example, the NBA's Western supporters evoked 'moral outrage and a sense of duty … to act on behalf of the Narmada people', while neglecting the ways in which the forces of capitalism emanating from the Global North were implicated in those injustices. These efforts were directed towards 'a localized, bounded community … on the basis of humanitarian concerns rather than emerging from global issues of interdependency [and] … a common struggle with the people of the Valley' (Shukla 2009, 141). We observe these very effects across different scales of advocacy within India – such as between the NBA's urban middle-class supporters and its local constituents. However, in a different political context, the NBA could have spawned into an alternative political project, one which would have focused on developing, rather than assuming as given, the principles and practices of 'equity, equality, participation, and ecological responsibility' (Sinha, Gururani, and Greenberg. 1997, 89–90).

The main lesson here is that instead of conceptualizing India's climate movement or climate justice movement as a monolithic phenomenon, it is important to investigate how diverse – and at times competing – frames and discourses of climate justice become part of climate governance debates in India. Careful scrutiny of environmental justice debates in India offers deep insights into the politics of competing frames. While indigenous rights are increasingly being recognized within the global community, some of India's prominent conservationists refuse to accept these arguments. Bittu Saigal, the editor of the *Sanctuary Asia* magazine, has referred to the enactment of the Forest Rights Act as Indian 'democracy's lowest hour' (Kashwan 2013). Shekar Dattatri, who gained fame via the Silent Valley movement, has referred to the forest rights movement as a means to 'grab land' (Dattatri 2019). According to two noted ecologists, Dattatri 'selectively mines the ecological literature' and engages in 'intentional obfuscation' to 'manufacture a perception' that the recognition of forestland rights contributes to forest degradation (Rai and Bawa

2019). The progress made by forest rights movements in India has, therefore, been undermined by a counter-movement run by India's elite conservationists.

India's climate justice movement will need to confront not just climate deniers and climate action delayers, but those who actively dismiss considerations of social justice in India's climate policy. Climate movements, including youth movements, need to engage with mass constituencies, learn from them, and support them. Social mobilization has proven most effective when it is structured as a process of engagement between a plurality of actors committed to the goals of strengthening state accountability and democratic governance (Kashwan 2017). Engaging with and strengthening domestic institutional arrangements to demand accountability of powerful market and state actors is not just a justice agenda, but a pre-requisite for effective climate action (Kashwan and Kodiveri 2021). This argument also resonates with scholars of international negotiation, who have argued that despite a longstanding focus on questions of international climate justice, India's climate policies do not demonstrate any serious appreciation of climate science, which would have brought to centre stage India's own climate vulnerabilities (Raghunandan 2019). While this chapter has focused mainly on domestic governance, transnational engagements can also be a fruitful avenue to bring about transformative change. The potential for such outcomes is enhanced when climate justice is 'reconceived as multiscalar with multiple entry points, and the interaction between scales [are] ... made explicit' (Fisher 2015).

The foregoing analysis has shown that realizing equity and justice are contingent on the processes used to manage such cross-scale coalition-building. Yet any such analyses and strategies must account for the formidable barrier that the broader political and economic context – including the nature of the Indian state – creates against the pursuit of social, environmental, and climate justice. Contemporary climate movements in India and elsewhere have an opportunity to learn from India's rich, albeit chequered, history of environmental movements.

References

Adve, N. 2019. 'How the Climate Justice Movement in South Asia Took a Big Step Forward Last Week'. *The Wire*, 26 September. https://thewire.in/environment/how-the-climate-justice-movement-in-south-asia-took-a-big-step-forward-last-week (accessed 26 December 2021).
———. 2021. 'Coming of Age of India's Youth Climate Movement'.
The India Forum, 24 March. https://www.theindiaforum.in/article/coming-age-india-s-youth-climate-movement (accessed 26 December 2021).
Agarwal, K. 2020. 'Citing Anti-Terror Law, Delhi Police Block Global Youth

Climate Activism Website'. *The Wire*, 23 July. https://thewire.in/environment/fridays-for-future-website-block-eia-prakash-javadekar-uapa (accessed 26 December 2021).

Banerjee, R. 2008. *Recovering the Lost Tongue: An Anarcho-Environmentalist Manifesto*. https://www.rahulbanerjeeactivist.in/manifesto.html (accessed 26 December 2021).

Baviskar, A. 2005. 'Red in Tooth and Claw? Looking for Class in Struggles over Nature'. In *Social Movements in India: Poverty, Power and Politics*, edited by R. Ray and M. F. Katzenstein. Lanham, MD: Rowman & Littlefield Publishers, Inc., 161–178.

———. 2019. *Uncivil City: Ecology, Equity and the Commons in Delhi*. New Delhi: Sage Publications and Yoda Press.

Bornstein, E. and A. Sharma. 2016. 'The Righteous and the Rightful: The Technomoral Politics of NGOs, Social Movements, and the State in India: The Righteous and the Rightful'. *American Ethnologist* 43 (1): 76–90. doi:10.1111/amet.12264.

Bosak, K. and K. Schroeder. 2004. 'Biodiversity Conservation and the Struggle for the Nanda Devi Biosphere Reserve'. *Focus on Geography* 48 (1): 1–6.

Caplins, L., S. J. Halvorson, and K. Bosak. 2018. 'Beyond Resistance: A Political Ecology of Cordyceps as Alpine Niche Product in the Garhwal, Indian Himalaya'. *Geoforum* 96: 298–308.

Dattatri, S. 2015. 'Silent Valley – A People's Movement that Saved a Forest'. *Conservation India*, 25 September. https://www.conservationindia.org/case-studies/silent-valley-a-peoples-movement-that-saved-a-forest (accessed 26 December 2021).

———.2019. 'How a Social Justice Tool Became a Means to Grab Land in India's Forests'. *Hindustan Times*, 2 July. https://www.hindustantimes.com/analysis/how-a-social-justice-tool-became-a-means-to-grab-land-in-india-s-forests/story-TPm9hWnFzRJavD1bKN2grM.html (accessed 26 December 2021).

Dogra, B. 2002. 'Whither the Chipko Years? The Fading Gains of Himalayan Conservation'. *India Together*, 5 June. http://www.indiatogether.org/environment/articles/postchipko.htm (accessed 26 December 2021).

Dubash N. K. 2019. *India in a Warming World: Integrating Climate Change and Development*. New Delhi: Oxford University Press.

Dwivedi, R.. 1998. 'Resisting Dams and "Development": Contemporary Significance of the Campaign against the Narmada Projects in India'. *European Journal of Development Research* 10 (2): 135–183. DOI: 10.1080/09578819808426721.

Fisher, S. 2015. 'The Emerging Geographies of Climate Justice'. *Geographical Journal* 181 (1): 73–82.

Ganesan, A. 2020. '*The Bastion* Dialogues: Dr. Amita Baviskar'. The Bastion, 9 April. https://thebastion.co.in/interviews/the-bastion-dialogues-dr-amita-baviskar/ (accessed 26 December 2021).

Ghosh, S. 2008. 'The Climate Crisis and People's Struggles'. *Mausam ... Talking Climate in Public Space* 1. http://www.thecornerhouse.org.uk/resources/results/mausam%20taxonomy%3A14 (accessed 26 December 2021).

Guha, R. 1989. 'Radical American Environmentalism and Wilderness Perservation: A Third World Critique'. *Environmental Ethics* 11 (1): 71–83.

Haque, M. S. 1997. 'Incongruity between Bureaucracy and Society in Developing Nations: A Critique'. *Peace and Change* 22 (4): 432–462.

Jain, S. 1984. 'Women and People's Ecological Movement: A Case Study of Women's Role in the Chipko Movement in Uttar Pradesh'. *Economic and Political Weekly* 19 (4): 1788–1794.

Jasanoff, S. 1993. 'India at the Crossroads in Global Environmental Policy'. *Global Environmental Change* 3 (1): 32–52.

Joshi, A. 2019. 'Politics of Environment: Language and India's Climate Change Movement'. *Livewire*, 22 November. https://livewire.thewire.in/politics/politics-of-environment-language-and-indias-climate-change-movement/ (accessed 26 December 2021).

Kashwan, P. 2013. 'The Politics of Rights-Based Approaches in Conservation'. *Land Use Policy* 31: 613–626. https://doi.org/10.1016/j.landusepol.2012.09.009.

———. 2014. 'Botched-up Development and Electoral Politics in India'. *Economic and Political Weekly* 49 (34): 48–55.

———. 2017. *Democracy in the Woods: Environmental Conservation and Social Justice in India, Tanzania, and Mexico.* Studies in Comparative Energy and Environmental Politics. New York: Oxford University Press.

Khastagir, N. 2002. 'The Human Face of Climate Change'. *Global Policy Forum*, 4 November. https://archive.globalpolicy.org/socecon/develop/2002/1104justice.htm (accessed 28 December 2021).

Koopmans, R. 2005. 'The Missing Link between Structure and Agency: Outline of an Evolutionary Approach to Social Movements'. *Mobilization: An International Quarterly* 10 (1): 19–33.

Lele, S. 2012. 'Climate Change and the Indian Environmental Movement'. In *Handbook of Climate Change and India: Development, Politics and Governance*, edited by Navroz Dubash. London and New York: Routledge, 208–217.

Li, T. M. 2014. 'Anthropological Engagements with Development'. *Anthropologie & développement* 37–38–39: 227–240.

Linkenbach, A. 2001. 'The Construction of Personhood: Two Life Stories from Garhwal'. *European Bulletin of Himalayan Research* 20 (1): 23–35.

Martinez-Alier, J. 2014. 'The Environmentalism of the Poor'. *Geoforum* 54: 239–241. https://doi.org/10.1016/j.geoforum.2013.04.019.

Mathews, R. D., S. Barria, and A. Roy. 2016. 'Up from Development: A Framework for Energy Transition in India'. Trade Unions for Energy Democracy, November. http://unionsforenergydemocracy.org/resources/tued-publications/tued-working-paper-8-up-from-development/ (accessed 15 August 2020).

Mathew-Shah, N. 2015. 'Disastrous Narmada Valley Projects: The Struggle to Resist Continues'. *Sierra Club*, 9 October. https://www.sierraclub.org/compass/2015/10/disastrous-narmada-valley-projects-struggle-resist-continues (accessed 28 December 2021).

Mawdsley, E. 2004. 'India's Middle Classes and the Environment'. *Development and Change* 35 (1): 79–103.

Mehta, L. 2013. *The Limits to Scarcity: Contesting the Politics of Allocation.* New York: Taylor & Francis.

Menon, R. V. G. 2012. 'Science for Social Revolution'. *Seminar* 637.

Mitra, A. 1993. 'There Can Be No Development without Women'. *Down To Earth*, 30 April. https://www.downtoearth.org.in/interviews/energy/there-can-be-no-development-without-women-30888 (accessed 28 December 2021).

News Minute. 2021. "'Still Fighting for Climate Justice": Disha Ravi Releases Powerful   Statement'. 13   March.   https://www.thenewsminute.com/article/still-fighting-climate-justice-disha-ravi-releases-powerful-statement-145184 (accessed 28 December 2021).

Omvedt, G. 1987. 'India's Green Movements'. Race and Class 28 (4): 29–38.

———.1997. 'Why Dalits Dislike Environmentalists'. *The Hindu*, 24 June. http://ces.iisc.ernet.in/hpg/envis/doc97html/envenv627.html (accessed 28 December 2021).

Purcell, M, and J. C. Brown. 2005. 'Against the Local Trap: Scale and the Study of Environment and Development'. *Progress in Development Studies* 5 (4): 279–297.

Raghunandan, D. 2019. 'India in International Climate Negotiations: Chequered Trajectory'. In *India in a Warming World: Integrating Climate Change and Development*, edited by N. K. Dubash. New Delhi: Oxford University Press, 187–204.

Rai, N. D. and K. S. Bawa. 2019. 'Giving Rights to Forest Dwellers will not Harm India's Forests'. *Hindustan Times*, 18 July. https://www.hindustantimes.com/analysis/giving-rights-to-forest-dwellers-will-not-harm-india-s-forests/story-7Ylyoz2rYPMpqQHqpFwrnWJ.html (accessed 28 December 2021).

Rajan, R. G. 2011. *Fault Lines: How Hidden Fractures Still Threaten the World Economy*. Princeton, NJ: Princeton University Press.

Rakesh, T. M. 2021. 'Lingering Effect of Disha Ravi Arrest'. *The Telegraph*, 25 February. https://www.telegraphindia.com/india/toolkit-case-lingering-effect-of-disha-ravi-arrest/cid/1807718 (accessed 28 December 2021).

Ramesh, J. 2010. 'The Two Cultures Revisited: The Environment–Development Debate in India'. *Economic and Political Weekly* 45 (42): 13–16.

———. 2015. *Green Signals: Ecology, Growth, and Democracy in India*. New Delhi: Oxford University Press.

———. 2017. *Indira Gandhi: A Life in Nature*. New Delhi, Simon and Schuster.

Rangan, H. 1997. 'Indian Environmentalism and the Question of the State: Problems and Prospects for Sustainable Development'. *Environment and Planning A* 29 (12): 2129–2143.

Rangan, H. 2000. Of *Myths and Movements: Rewriting Chipko into Himalayan History*. London, New York: VERSO.

Ray, R. and M. F. Katzenstein. 2005. *Social Movements in India: Poverty, Power, and Politics, Asia/Pacific/Perspectives*. Lanham, MD: Rowman & Littlefield.

Routledge, P. 2003. 'Voices of the Dammed: Discursive Resistance Amidst Erasure in the Narmada Valley, India'. *Political Geography* 22 (3): 243–270.

Satheesh, S. 2019. 'Hundreds of India Villages under Water as Narmada Dam Level Rises. *Aljazeera*, 23 September. https://www.aljazeera.com/news/2019/9/23/hundreds-of-india-villages-under-water-as-narmada-dam-level-rises (accessed 28 December 2021).

Shah, E., J. Vos, G. J. Veldwisch, R. Boelens, and B. Duarte-Abadía. 2019. 'Environmental Justice Movements in Globalising Networks: A Critical Discussion on Social Resistance against Large Dams'. *Journal of Peasant Studies* 48 (5): 1008–1032.

Shaji, K. A. 2015. A Requiem to the Kunthipuzha. *The Hindu*, 14 December. https://www.thehindu.com/news/national/kerala/a-requiem-to-the-kunthipuzha/article7985473.ece (accessed 28 December 2021).

Shukla, N. 2009. 'Power, Discourse, and Learning Global Citizenship: A Case Study of International NGOs and a Grassroots Movement in the Narmada Valley, India'. *Education, Citizenship and Social Justice* 4 (2): 133–147.

Sinha, S. 2021. '"Strong Leaders", Authoritarian Populism and Indian Developmentalism: The Modi Moment in Historical Context'. *Geoforum* 124: 320–333.

Sinha, S., S. Gururani, and B. Greenberg. 1997. 'The "New Traditionalist" Discourse of Indian Environmentalism'. *Journal of Peasant Studies* 24 (3): 65–99.

Sud, N. 2020. 'The Actual Gujarat Model: Authoritarianism, Capitalism, Hindu Nationalism and Populism in the Time of Modi'. *Journal of Contemporary Asia*. DOI: 10.1080/00472336.2020.1846205.

Swain, A. 1997. 'Democratic Consolidation? Environmental Movements in India'. *Asian Survey* 37 (9): 818–832.

Swarnakar, P. 2019. 'Climate Change, Civil Society, and Social Movement in India'. In *India in a Warming World: Integrating Climate Change and Development*, edited by N. K. Dubash. New Delhi: Oxford University Press, 253–272.

The Wire. 2021. 'More Than 400 Academics, Activists Condemn Arrests of Disha Ravi, Nodeep Kaur, Others'. 22 February. https://thewire.in/rights/academics-condemn-disha-ravi-nodeep-kaur-arrests (accessed 28 December 2021).

UNFCCC. 2007. 'The Report of the Conference of the Parties on Its Thirteenth Session, Bali, Indonesia, 3–15 December 2007. Addendum, Part Two: Action Taken by the Conference of the Parties at Its Thirteenth Session'. Decision 2/CP.13. https://unfccc.int/documents/5079#beg (accessed 28 December 2021).

Vombatkere, S. G. 2016. 'Narmada Bachao Andolan: Thirty Years of Resistance and Reconstruction'. *Mainstream Weekly*, 9 September. http://www.mainstreamweekly.net/article6663.html (accessed 28 December 2021).

Warrier, S. G. 2018. 'Silent Valley: A Controversy That Focused Global Attention on a Rainforest 40 Years Ago'. *Mongabay*, 1 February. https://india.mongabay.com/2018/02/silent-valley-a-controversy-that-focused-global-attention-on-a-rainforest-40-years-ago/ (accessed 28 December 2021).

Wessler, H. and R. Freudenthaler. 2018. 'Public Sphere'. *Oxford Bibliographies: Authority and Innovation for Research: Communication*. https://doi.org/10.1093/obo/9780199756841-0030.

The House with No Windows

Interviewee 10 has a joint family of 9
10 people live in a thatched house
At the border of Karnataka and
Tamil Nadu
The thatched house has two rooms
A kitchen and a room of everything else
The house is dark even in the daytime
Gaps in the thatched roof
Send sharp streaks of light
Cutting the air like a knife
His wife is applying cow dung on
the walls
Interviewee 10 takes me to the house
of his best friend
'The house of seven rooms,' he says
An enclosed courtyard opens to the sky
Awash with sunlight
We are not welcomed inside
We sit on the inviting veranda
And talk about the potted plants in the
14 windows
The talk turns to their friendship
The best friend says,
'We are best friends, but he can never enter
my house. Never! We are different'
—Praneeta Mudaliar

This excerpt from Praneeta Mudaliar's poem captures a fundamental truth about India. Caste is omnipresent—it shapes how social, cultural, economic, political, legal, and educational systems work. Even more importantly, caste is inscribed in our physical infrastructure and geography. It determines which places one is permited to enter or not. The social structures of caste, along with class and gender, also influence whether the sun brings joy in an open courtyard or whether it is a source of heatstroke from working long hours on a construction site. Yet, for some reason, caste is invisible in much of the research on the environment and climate crisis in India.

The rules, norms, and conventions that underlie the caste system may be less visible in some places and at some times, but they have not weakened. The variegated nature of India's caste hierarchy makes it infinitely more complex, much older, and more deeply ingrained in our lives than, say, the race system in the United States. Trauma runs through the lives of Dalits and other marginalized groups, as invisibly and as constantly as blood flows in our veins; it is part of who we are. As the worsening impacts of the climate crisis decimate our infrastructure and drown us in our filth, it will further aggravate the repressions and brutalities of the caste system.

CHAPTER 8

Reimagining Climate Justice
as Caste Justice

Srilata Sircar

The contribution of colonialism and imperial expropriation to the unfolding climate crisis has been well documented on a global scale. This chapter seeks to interrogate the role of caste as a structural element in shaping environmental inequities within India and beyond. Scientists across disciplines agree that the current system of production is unsustainable at the planetary level, even if a consensus on how to address this issue remains elusive. I argue that in the case of India, accounting for historical and contemporary caste-based extraction is crucial for any meaningful realization of climate justice.

Globally, academic scholarship and policy have come to acknowledge the uneven and unjust ways in which the burden and responsibility for the current crisis are distributed across nations, ethnicities, races, and genders. There is an emerging consensus that the historical pathways of colonialism and capitalist development are directly responsible for this uneven distribution. This pattern is seen across the histories of energy production, plantation economies, and commercial agriculture, as demonstrated in the detailed work of political ecologists (for example, Li 2017). Consequently, the idea that mitigation, adaptation, and resilience-building strategies must account for this historical unevenness is no longer controversial.

We see this acknowledgement in the principle of 'common but differentiated responsibility' formally adopted by the United Nations in 1992. Under this principle, world governments recognize the lesser contribution of formerly colonized countries such as India towards planetary environmental degradation. This can be read as an acknowledgement of the unequal distribution of political power and economic prosperity across world nations because of colonialism. Acknowledging this historicity of the climate crisis is important, but our understanding of it would remain incomplete without a serious stock-taking of those dimensions of inequality and unevenness that significantly pre-date the rise of colonial capitalism and are yet implicated in its development trajectory. These dimensions of inequality often operate at the national or sub-national levels and therefore escape scrutiny on the global stage. In the case of India, one such important and all-pervasive dimension of inequality is caste.

For decades, anthropological and historical scholarship on caste focused only on ritual, scriptural, and mythical dimensions, thus constructing the issue as a matter of religion alone. Anti-caste scholars and activists such as Ambedkar, Phule, and Periyar have resisted such 'orientalist' representations of caste. One main aim of anti-caste scholarship has been to expose how caste operates as a spatially organized institution that is directly connected to economic and other material resources (Thorat and Newman 2010). Despite this long tradition of anti-caste scholarship, research on the material and structural dimensions of caste has remained limited. While it is empirically well documented that access to land, labour conditions, inheritance of wealth, and opportunities for social mobility are all deeply differentiated along caste lines, these arguments are yet to be incorporated into scholarship on environmental and climate justice (for example, Vijayabaskar and Wyatt 2013).

In this chapter, I demonstrate how in the case of India, pre-existing social relations determined by caste have shaped capitalist development in the region, which in turn has influenced the trajectory of the climate crisis. Thus, one's contribution to the climate crisis and exposure to its effects are related closely to one's caste location in Indian society. This is why I argue that the question of climate justice in India is inseparable from the question of caste justice. In other words, climate justice *is* caste justice.

To elucidate this argument, I draw from historical and contemporary empirical evidence. I present the argument in four sections. The first section outlines how caste and capital co-evolve in the context of colonialism. The discussion on colonial property regimes shows how caste ideology reinforced capitalist frameworks to produce displacement, dispossession, and extraction. The next section demonstrates how historical caste-based displacement continues to shape land relations, labour relations, and resource use in urban India. I use examples from urban housing to establish the continuities between colonial policies and neoliberal reforms. The third section discusses current policy dispositions towards tackling urban sanitation issues and how they reproduce unequal caste relations and further entrench caste power through the deployment of digital governance. The final section highlights the chasm between anti-caste politics and urban environmentalist agendas. Throughout these four sections, I foreground the need to recognize the social reality of caste to address its distributional and procedural outcomes. Based on these arguments, the conclusion will set an agenda for anti-caste climate activism and scholarship going forward. As in the rest of this volume, this chapter, too, will use the vocabulary of 'upper' and 'lower' castes as shorthand to refer to the relative position of groups in the social hierarchy and power structure of caste. The quotation marks here indicate my personal disavowal of this system of caste hierarchy. The usage here is purely descriptive and not normative.

Caste, capitalist development, and colonial property regimes

Several studies have documented the role of caste in perpetuating inequality in India (Thorat and Newman 2010). Traditionally, the caste structure was maintained through hereditary occupations and by enforcing strict rules of endogamy. Caste elites traditionally enforce these rules through social sanction and ostracization. The abolition of egregious caste practices such as untouchability (as mandated by the Indian Constitution), and the opening up of higher education and employment opportunities to all caste groups, were welcome steps but they did not adequately address centuries of deprivation. For instance, the 70th round of the National Sample Survey in India found that more than 70 per cent of all farmers from the lowest caste group (Scheduled Castes) worked as agricultural labourers dependent on daily or seasonal wages from upper caste landlords. The survey further found that close to 60 per cent of all rural Scheduled Caste households were landless and entirely dependent on casual wage work (*Hindustan Times* 2018).

A study of the occupational profiles of caste groups showed that traditionally oppressed groups such as Dalits and Adivasis are overwhelmingly over-represented in the informal sector where wages are lower, work conditions more precarious, and social security non-existent (Singh and Thorat 2014). While the study found that, by and large, occupational mobility has indeed improved for elite and mid-ranking caste groups in the postcolonial period, for the lowest-ranked groups, that is, the 'ex-untouchables', it has remained nearly impossible. The starkest instance of this can be seen in the case of 'manual scavenging' – the manual, unmechanized, and unprotected cleaning of dry latrines, sewers, drains, septic tanks, and railway tracks. Even though the practice was declared illegal in 1950, workers are hired for manual scavenging by public and private actors alike. According to estimates by Safai Karmachari Andolan, an activist movement aimed at eliminating manual scavenging, approximately 98 per cent of all workers employed in this kind of work are Dalits and predominantly women (Safari Karmachari Andolan n.d.).

How is it that manual scavenging not only continues even after seven decades of independence and affirmative action but is still performed only by a specific caste group? Part of the answer is in how public institutions tap into the caste order and its hereditarily assigned occupational roles. Ambedkar described it in *Annihilation of Caste* (2004 [1944], 4.1) as 'not merely a division of labour … [but] … also a division of labourers'. He critiqued this division of labourers for its rigid hierarchy, denial of agency to those it stratifies, and obstruction of opportunities for genuine solidarity and nation-building. He defines caste society as a 'society in which some men are forced to accept from others the purposes which control their conduct' (Ambedkar 2004 [1944], 14.4). This fundamental undermining of agency and dignity by caste has been obfuscated in the way the institution has come to be codified under colonial rule.

The administrative categories through which caste is made formally legible and measurable in society today – that is, Scheduled Caste, Scheduled Tribe, and Other Backward Classes – are an inheritance from colonial rule. Each of these categories stands in for serious deprivations and historical injustices; however, in policy discourse, they are presented, especially to the global community, using the vocabulary of socioeconomic and educational metrics. These administrative categories obscure more than they reveal. For one, each of them encompasses hundreds, if not thousands, of sub-categories of *jatis* or *janjatis*, each of which is a community with its own unique lived experience of caste. More importantly, the categories privilege the so-called higher castes by defining them as the 'general' category while othering the so-called middle and lower castes via various labels that carry negative societal connotations. By doing so, they obscure the true nature of caste – a system of oppression that devalues and demeans the very existence of those it marginalizes. These categories do not expose and make visible the operation of caste power and those who have benefitted from it. What is lost in this partial and aggregated reading of caste is the long and complex history of how caste and colonialism co-produced capitalist development, whose effects have since compounded into the current climate crisis that presents inordinate threats for the oppressed. To understand the relationship between climate justice and caste justice, we must recognize and reconstruct this history of collusion.

One protagonist in this story of capitalist development is the idea of property. Postcolonial historians have illustrated how colonial powers turned occupied land into 'property' (Bhandar 2019). By doing so, the colonial state succeeded in generating value for itself through rent, taxation, and claims to the produce of the land. In British India, the colonial government established standardized property regimes to exploit the environment and local population to extract the maximum possible value. This was accomplished by solidifying existing caste relations by embroiling them in these property regimes. Bhambra has argued that colonialism provided the foundational structure for the development of global capitalism (Bhambra 2020). The value extracted in the form of revenue from colonies was poured into creating and sustaining the core infrastructure of the capitalist economy. This was also seen in the Americas, where the commodification of enslaved labour and resultant conversion of people into property provided the foundation for the emergent capitalist system (Robinson and Quan, 2019). However, a similar reckoning of how property contributed to global capitalism in colonial South Asia is lacking. Any such reckoning must include caste as a central analytical category. To understand this process, we now turn to two historical examples from the strategic and high-value agricultural provinces of Bengal and Punjab.

In Bengal, one of the most far-reaching colonial interventions was the Permanent Settlement of 1793 (Guha 1982). This act created an institutionalized property regime whereby the colonial state assumed the role of a landlord. The exact contours of the social change brought about by the Permanent Settlement are still being debated by historians, but two things remain undisputed. First, the creation of legal land titles established a land market that benefitted upper caste groups, both as sellers in rural areas and buyers in urban areas. Second, the formalization of land titles and revenue extraction processes led to increased coercion and exploitation of landless lower caste groups. In this colonial property regime, those with property rights could easily pay for their dues to the colonial state by either using extractive force or by selling off their land titles and revenue rights to the highest bidder. However, the actual tillers remained tied to their inherited status with no formal rights, while bearing the onus of having to produce more and more as the demand for revenue increased (Ray 1974).

A similar process unfolded in the province of Punjab. Drawing on the lessons learned from their experience in Bengal, colonial administrators sought to make the rural context legible to the state through a homogenous system of classification. This paved the way for 'improvement' planning. The colonial state undertook the humongous administrative task of mapping and designating all the available land within the province into categories that were legible to it. Until the 1870s, only about 40 per cent of the surveyed area in Punjab was under settled cultivation; under the colonial administration, the share of this category increased significantly. To do this, the colonial state made all peasants knowable and countable and turned all available land into productive assets, that is, property. Bhattacharya refers to this as the 'great agrarian conquest', which led to the slow but significant erasure of nomadic and pastoralist ways of life, a wide gamut of common property rights, and seasonal rights to resources such as wells and forests (Bhattacharya 2019). In the words of a colonial chief commissioner of Punjab, the administrators conducting the mapping and settlement 'disposed of at least 80,000 petty rent-free tenures' and 'decided some 6000 suits to landed property or ancestral rights' (Bhattacharya 2019, 76). In making these decisions, the colonial administrators identified the dominant caste in each part of the province and formalized their customary practices through the use of *bhaichara* tenures or rights of ownership based on fraternity.

As in the case of Bengal, the colonial settlement institutionalized only two kinds of land relations – ownership and tenancy. Through the use of *bhaichara* tenures, the land rights of upper castes were legally secured as ownership and those who were not considered to be part of the *bhaichara* fraternal community were automatically accorded tenant status irrespective of their actual use of and relationship to the land (Bhattacharya 2019). Lower castes, non-agricultural castes, and women were bereft of land rights. In cases involving especially complex customary land rights, ownership and tenancy were sub-categorized into superior and inferior, with each accorded varying degrees of rights and protection. But on the whole, land rights, including the customary claims of the lower castes and Adivasis, went entirely unheeded. Similar settlement interventions were conducted across the whole of British India. This history of dispossession and revenue extraction directly connects caste relations in the subcontinent to capitalist development, the ongoing climate crisis, and the question of climate injustice.

The decades following the settlement of the Bengal province saw a massive rise in revenue collections – within the first three decades, there was an increase (Roy 2013). This revenue was channelled into private profits and British public sector works, including the construction of infrastructure in other colonies such as roads, railways, and factories in the Americas (Patnaik 2017). As argued by Bhambra, these investments formed the bedrock of the core infrastructure of the colonial economy and therefore global capitalism (Bhambra 2020). Thus, the history of caste injustices is intricately connected to climate injustices, not only within India and South Asia but also in global dialogue on tackling the climate crisis. The principle of common but differentiated responsibility must therefore be extended to account for caste-based domestic inequalities and be concretely reflected in climate policy at both the international and national levels. In the next section, we look at how this caste-based regime of property, which was reinforced under colonial rule, continues to shape everyday life in postcolonial India.

The everyday contours of caste in contemporary India

The institutionalization of property and reinforcement of the material basis of caste have rendered social groups designated as lower castes disposable and have normalized their displacement. This continues to be the norm across different geographies even in contemporary times. In this section, we consider how this plays out in urban contexts and especially in everyday land and labour relations. This section demonstrates how this structural disposability of the so-called lower castes relates to the distributional, procedural, and recognitional aspects of climate justice in the Indian context.

In the light of worsening climate risks, migration is often cited as a valid and appropriate adaptation strategy to cope with environmental, socioeconomic, and political stress (Adger and Adams 2013). India's seasonal and circular labour migrants have used this strategy for decades. However, they continue to face exclusion and displacement within their host cities, which further heightens their vulnerability to extreme weather events. Both ethnographic and statistical studies show that the deprived caste groups – Scheduled Castes, Scheduled Tribes, and Other Backward Classes – make up the majority of the circular migrant population (Deshingkar and Akter 2009). Though there is no disaggregated data on the migration practices of specific caste groups, ethnographic evidence has shown how the absence of land rights, lack of security of tenure, and stagnant agricultural wages have pushed subaltern caste groups made up of small cultivators and landless wage workers to migrate to urban areas in search of informal employment (Breman 1996; Harriss-White 2003; Sircar 2018).

The mobile informal workforce that urban India is dependent on consists of the very same dispossessed and displaced caste groups. The landlessness and precarity of tenure institutionalized under the colonial regime were further entrenched by the failure of redistributive land reforms in postcolonial India (Kashwan 2017). The introduction of neoliberal reforms in the 1990s firmly established metropolitan urban centres as the drivers of growth – these cities in turn became dependent on this massive itinerant workforce.

There are an estimated 100 million seasonal and circular migrants working in India's informal economy. The informal sector accounts for more than 95 per cent of all employment in the country and relies heavily on migrant workers (Deshingkar and Akter 2009). Of these, about 15 million are estimated to be child migrants. Labour migration has been a key structural element of the Indian economy for several decades, but it is yet to be given due attention in policy. This was made most starkly visible during the COVID-19 lockdown of 2020 when millions of migrant workers journeyed home to their villages by foot, bicycle, and other ad-hoc modes of transport. Considered the largest exodus since the Partition in 1947, images of millions of people walking down empty highways occupied headlines for weeks. While the pandemic and ensuing lockdown have no doubt worsened the already precarious situation of these migrant workers, this presents an opportunity to investigate their marginalization even during relatively prosperous times. It is also important to examine how historic caste-based deprivations feed into and reinforce this marginalization. Throughout the postcolonial period, and especially since the economic liberalization of the 1990s, cities across India have adopted policies that push migrant workers to live in under-served urban settlements or 'slums' that are exposed to greater climate risk. One of the common mechanisms through which this takes place is the removal of informal settlements from urban land that is considered to be high value (and therefore up for 'improvement') by blaming them for environmental degradation (Baviskar 2020). The discourse that planners, municipal authorities, caste elites, and judicial institutions mobilize to justify these displacements is reinforced by both overt and covert caste prejudices. Thus, caste operates as a key organizing factor in the distributional and procedural aspects of urban climate governance.

In Delhi, the post-liberalization period saw the emergence of 'green speak' in urban planning, which envisioned the environment as an aesthetic category within a 'slum-free' and 'world class' city (Ghertner 2011). In this vision, the informal settlements of migrant workers were seen as inherently polluting because of their aesthetic departure from the desired urban vista. Although caste is seldom named or directly evoked in this discursive violence, its reliance on caste ideology is all too evident.

A case in point is the Pushta *basti* settlement on the bank of the Yamuna in Delhi, where most residents are from the so-called lower caste communities. The settlement first emerged in the 1970s and grew exponentially in the run-up to the 1982 Asian Games when construction workers were brought in to build the infrastructure for this high-profile event (Bhan 2017). Since the 1990s, a series of public interest litigations (PILs) have been filed by private actors seeking to remove Pushta *basti* and displace its residents. These include owners of industrial units and resident welfare associations of upscale neighbourhoods that release their unprocessed effluents and sewage into the river, respectively. By the Delhi Water Board's own admission, a majority of the river's pollution can be traced to these elite residential and industrial units rather than to the *basti* (Ghertner 2011). In a 1994 petition to the Delhi High Court, a group of factory owners demanded that the municipal authorities 'destroy infectious huts and shed[s] in order to prevent the spread of any dangerous diseases' (Ghertner 2011, 145). The court upheld this claim and ordered the removal of the settlement. This labelling of lower caste settlements as inherently infectious is steeped in the caste ideology of purity and pollution.

In the aftermath of the Bangalore plague of 1898, the colonial administration represented the epidemic as 'a disease of locality' with origins in the 'cultural pollution' of those deemed unsanitary (Ranganathan 2018, 1391). Even after it was proven that the plague was transmitted by rat fleas and not cultural practices, the municipality continued to focus on aesthetic interventions in the White and elite native parts of the city instead of rebuilding the crumbling sanitation infrastructure in the so-called infectious localities. This pattern has continued into contemporary times. Björkman's (2015) study of piped water provisioning in Mumbai shows that elite residential complexes, even when constructed using dubious means and with alterations outside the master plan, receive water connections and regular supply. In stark contrast, settlements designated as 'slums' due to their working-class/caste character, even when constructed within the framework of planned development, are displaced to marshlands in the city's fringes and deprived of basic amenities. In these measures undertaken by urban planning agencies, it is easy to see how the distribution of blame, responsibility, and risk is unfairly skewed against the caste subaltern.

Resettlement procedures are similarly influenced by caste ideology. In 2002, Ambedkar Slum Utthan Sangathan (ASUS) – a coalition of displaced households seeking fair resettlement in Delhi – moved the Delhi High Court against the municipality for assigning them resettlement flats of 24 square meters as opposed to the 60 square meters that they had been originally promised 30 years prior. The court declined to stop their forced resettlement after a court-appointed commissioner found the flats to be 'commensurate with the status of the persons sought to be shifted' (Bhan 2017, 465). This issue of 'status' is, of course, laden with caste prejudice. The same 24 square meters would be deemed inadequate for an elite caste/class family but was deemed suitable for displaced migrant workers from marginalized castes. Thus, the caste identity of migrant workers is invoked as a marker of 'status' to justify depriving them of their right to property and fair housing. Municipalities and urban development authorities routinely use the absence of land titles and formal ownership rights to justify dispossessing de facto users of urban land, even if many of these groups had previously negotiated with the bureaucracy to secure basic services in the form of water and electricity meters.

Somewhat ironically, PILs have emerged as a tool for elite urban residents to displace subalterns from both urban spaces and the imagination of urban citizenship. In this process, the idea of property features as a key determinant of social and legal status. In an earlier 2002 judgment, the Delhi High Court had considered the welfare of property-owning elites as 'public interest' while failing to similarly uphold the rights of property-less members of the 'public' (Bhan 2016). This logic is replicated across all urban development interventions in the age of neoliberal reforms. For instance, under the 100 Smart Cities Mission, municipalities and smart city authorities across India are required to facilitate land acquisition for city extension and the building of 'satellite towns' beyond municipal boundaries. In Nashik, Maharashtra, more than 700 acres of agricultural land is being acquired for 'greenfield development', requiring a consensus among more than 300 farmers (Pawar 2019). The land in concern is currently being used to cultivate grapes – a lucrative cash crop – and offers employment to landless wage workers from the Dalit and Adivasi communities. These workers are not recognized as stakeholders in the negotiations even though they will lose their livelihoods if the project moves forward (Smart Cities Council 2020a).

Similar examples abound across small and large urban centres. In the city of Shimla – a part of the Punjab province under colonial rule and the summer capital of British India – the main smart city intervention was the redevelopment of the main commercial zone, Lower Bazaar. The proposed smart city plan aims to promote adaptation and resilience building by improving sewage, drainage, and solid waste management infrastructure (Smart Cities Council 2020b). However, it fails to recognize the residents of Krishna Nagar – a settlement downhill from Lower Bazaar – as equal stakeholders in this redevelopment. The settlement in Krishna Nagar is located on top of soil deposited from mountain excavations to build the city centre in the colonial era. This makes the settlement structurally prone to landslides and flooding. While the first houses were constructed by migrant labourers from the surrounding plains, later residents were sanitation workers employed by the municipality. They were provided government staff quarters constructed by the municipality, but which are now referred to as a 'slum' due to lack of upkeep and recurrent damage from flooding (Datta 2019).

The tendency to disregard formality and legality and use the denomination 'slum' to refer to any settlement that is deemed displaceable because of its aesthetic and the social status of its residents is in fact officially validated. The 1971 Maharashtra Slum Areas (Improvement, Clearance and Redevelopment) Act allows any area to be designated as a 'slum' if it 'is or may be a source of danger to the health, safety or convenience of the public of that area or of its neighbourhood, by reason of the area having inadequate or no basic amenities' (Björkman 2015, 101). Thus, the municipality's failure to extend basic services is obscured by mobilizing the casteist discourse of hygiene and safety. Subaltern caste groups are hence marginalized within urban housing through the following interlinked mechanisms: (a) the erosion of customary rights through the discursive mobilization of casteist tropes, (b) systematic denial of formal property ownership, and (c) designation of property ownership as the sole means to securing participation in decision-making.

This is evident in Chu and Michael's research in the cities of Bengaluru and Surat. The authors argue that migrant workers in the informal sector 'embody intersecting forms of environmental marginality' that are based on existing structures of social power such as gender and caste (Chu and Michael 2019, 152). Recognizing this social reality and the deprivations that result from it is a necessary precursor to any meaningful climate justice intervention. I have extended this argument to demonstrate the logical continuity between colonial-era policies and contemporary neoliberal interventions. Both benefit from the existence of caste and reinforce caste ideology to further entrench material inequalities. In the following section, we will take a closer look at the urban sanitation sector, which exemplifies the intersection of caste-based extraction of labour and urban climate governance.

Caste, urban sanitation, and digital governance

In recent years, urban sanitation has emerged as a prominent sector for climate governance and sustainability interventions. Recurrent urban flooding in high-profile metropolitan centres such as Mumbai (2005), Chennai (2015), Kochi (2018), and Bangalore (2020) has attracted significant attention. Two highly publicized policy regimes have been the key drivers of interventions in urban sanitation work – the Swachh Bharat Mission and the 100 Smart Cities Mission. In this section, we look at some of the interventions undertaken as part of these missions and dissect them from the perspective of caste justice. The analysis shows that these policy approaches are missed opportunities to recognize caste injustice.

Both of these policy regimes place sustainability at the centre of their agendas. The Swachh Bharat Mission, while being predominantly focused on rural India, aspires to attaining 'open defecation free' (ODF) status and 'universal sanitation' through 'cost effective and appropriate technologies for ecologically safe and sustainable sanitation' (Swachh Bharat Mission – Grameen n.d.). On a similar note, the Smart Cities Mission seeks to create 'cities that provide core infrastructure and give a decent quality of life to its citizens, a clean and sustainable environment and application of "Smart" Solutions' (Ministry of Urban Development 2015, 5). The mission guidelines identify 'sanitation, including solid waste management' as a core infrastructure that all smart cities must address. While retrofitting and integrating digital technology into urban governance is a key element of the mission, a thematic mapping of the proposed smart city projects reveals that urban renewal, redevelopment projects, and physical infrastructure with the potential to generate high revenue constitute much of the 'smartness' (Taraporevala 2018). Here, we see a clear continuity with the urban development approach of earlier decades, which relies on the displacement and disposability of caste subalterns. As the following discussion shows, by applying the same exclusionary model of urban planning, the mission fails to address the issue of caste that underlies the sanitation crisis in urban India.

Urban infrastructure, much like the rural property regime, has been foundational to the creation of the contemporary land and market regimes that fuel capitalism and, by extension, the climate crisis (Ramesh and Raveendranathan 2020). As far as sanitation goes, both urban and rural infrastructure are entirely reliant on deeply exploitative labour extraction of Dalit workers. The most grotesque form of this is the outlawed practice of 'manual scavenging'. In the Employment of Manual Scavengers and Construction of Dry Latrines (Prohibition) Act of 1993, manual scavenging is defined as the practice of manually handling human excreta for the purpose of cleaning a latrine, pit, or drain. However, Shankar and Swaroop (2021) argue that there are many other forms of sanitation work in which human beings – overwhelmingly Dalits – are forced to come into contact with human excreta. These include cleaning railway stations and tracks as trains in India discharge sewage material directly onto the tracks, cleaning storm drains carrying sewage instead of stormwater, and cleaning septic tanks in private and public buildings.[1]

According to the Ministry of Social Justice and Empowerment, there were 282 recorded deaths from manual scavenging between 2016 and 2019 (Desai 2020). Safai Karamchari Andolan estimates that there are 1.2 million people involved in manual scavenging (Safai Karmachari Andolan n.d.). Other studies have estimated that those involved in the manual cleaning of drains and septic tanks have a substantially lower life expectancy than average (Das 2018). Due to the paltry wages offered to manual scavengers, the absence of any protective gear or life insurance, the lack of agency in determining working conditions, and the enduring caste character of this exploitative system, Shankar and Swaroop (2021) argue that this practice constitutes slave labour and the ensuing violence against Dalits is of genocidal proportions.

Yet, the two dominant policy regimes, the Swachh Bharat Mission and Smart Cities Mission, make no mention of caste exploitation in sanitation work and articulate no vision for tackling this mammoth problem. In fact, the Swachh Bharat Mission actively builds on this horrific legacy. As part of this mission, more than 5.5 million new toilets have been constructed that are reliant on pits or septic tanks that are not necessarily linked to a drainage or solid waste management system network (Das 2018). While the mission celebrates the number of new toilets constructed, it does not lay out any concrete plans for expanding faecal sludge management infrastructure to cope with this massive volume of sewage. The responsibility for this is passed on to village- and city-level authorities without any system established for monitoring or support. This is where the role of municipalities becomes

---

[1] In 2019–2020, a large number of train coaches were fitted with bio toilets, which addresses the problem to some extent by processing human waste into organic matter. However, it does not address the caste character of sanitation work.

relevant. Under the Smart Cities Mission, municipalities and smart city cells have the opportunity to upgrade and modernize sanitation infrastructure to eliminate the reliance on caste-based exploitation. However, this has not been the case. By and large, cities have opted for superficial beautification projects under the Smart Cities scheme, introducing digital components into existing infrastructure but not fundamentally improving or rehauling them (Khan, Taraporevala, and Zérah 2018).

While surveillance is quite central to the notion of 'smartness' in these projects, it has very different implications for different groups of city residents (Monahan 2018). Many cities have introduced digital technology into sanitation management, which some have referred to as a 'surveillance revolution'; however, this surveillance primarily targets Dalit sanitation workers (Khaira 2020). In many of the smart cities, such as Nagpur, Nashik, Pimpri Chinchwad, Bengaluru, Trichy, Vishakhapatnam, and Patna, one of the first smart city projects to be implemented was the addition of GPS trackers to waste collection carts and the vehicles used by sanitation workers (*Moneycontrol News* 2020). In other places such as Chandigarh and Panchkula, sanitation workers themselves have been asked to wear smartwatches with GPS tracking. The data from these devices are tracked by municipalities and urban managers with the intention of disciplining workers with pay cuts and other punitive measures for any perceived shortcoming. At the same time, urban policy programmes have failed to increase the budget allocated for waste processing and sanitation infrastructure. As mentioned before, sanitation work in both public and private spaces continues to be dangerous, insecure, and life-threatening and is performed predominantly by Dalit workers. This reflects the colonial approach of turning each individual into a mapped and legible entity for the purpose of profit maximization but without any corresponding investments in their welfare. It is thus clear that the postcolonial urban regime in India, especially under neoliberalism, has been replicating the same logics as the colonial state while treating caste subalterns as displaceable and disposable.

It is evident that urban life in India is maintained through caste-based exploitation. The narrative of caste in urban India would, however, be incomplete without a mention of the strong and resilient resistance movements led by anti-caste activists and caste subalterns. In the next section, we will trace the contributions of anti-caste politics towards reimagining the urban sustainability agenda. Through this, I seek to demonstrate that the mainstream climate justice movement in India is yet to address caste seriously and that doing so is imperative for realizing a future course of action.

Over many past decades, anti-caste social movements have been striving to expose the close connections between caste injustice and resource extraction. Examples of this abound in agrarian and forest-based contexts. Protest movements against development-induced displacement, struggles for institutionalizing the rights of forest-dwelling communities, and peasants' movements for land rights and sustainable agriculture practices all have a marked anti-caste character, often articulated through the vocabulary of human rights (Ranjan and Kashwan 2021). These movements highlight distributional disparities in terms of which resources are seen as available for extraction and who is left to shoulder the associated risks. They also expose procedural disparities in terms of who gets to make decisions about resource use and whose lives are most affected by these decisions. The issue of recognition has been raised by some recent anti-caste movements such as the Dalit protests in Una, Gujarat, in 2016. Following a horrific attack on four Dalit men by upper caste cow-protection vigilante groups, Dalit groups across many districts went on a strike and refused to engage in their inherited occupation of handling animal carcasses. The protests involved collective action with a visceral impact such as letting animal carcasses pile up in villages, dumping carcasses outside the district collector's office, and boycotting Independence Day celebrations (Thekaekara 2016). The movement foregrounded the indispensable role played by Dalits in sanitation work and the integral connection between their labour and the natural and built environments. The movement was successful to the extent that one of its leading spokespersons, Jignesh Mevani, got elected to the Gujarat State Assembly in 2017. The movement's attention expanded from opposing atrocities to claiming land rights and securing land titles for Dalits (Dalit Camera 2016).

Movements such as the Una anti-caste uprising and Safai Karamchari Andolan are left out of the mainstream imagination of climate justice in urban India. Many urban groups have successfully challenged development interventions based on environmental concerns, but they largely remain confined to what Amita Baviskar (2020) describes as 'bourgeois environmentalism'. This is a form of environmental activism that is targeted toward establishing an aesthetic of ordered spaces, where nature is available for elite recreation, and the blame for environmental degradation is placed on caste/class subalterns. Institutionalized forms of bourgeois environmentalism can be seen in the functioning of agencies such as the National Green Tribunal (NGT), the smart city cells, and municipalities. One of the mechanisms through which this takes place is the procedural exclusion of caste/class subalterns from decision-making. Two examples discussed below elucidate this.

In 2017, the NGT banned construction in the old town area of central Shimla, essentially blocking all redevelopment proposed under the Shimla Smart City plan (Sharma 2019). It upheld the order in 2018, quashing the review appeal of the state government. The campaign that had brought the issue before the NGT was led by a group of elite environmentalists who were justifiably concerned about the sensitive ecological context of the city and the stress the proposed redevelopment would cause it. However, neither their campaign nor the NGT order took any cognisance of the community already living in ecological duress in the Krishna Nagar area. None of the residents of this area was consulted, and even after the NGT ban on construction, they continued to live in unsanitary conditions and at risk of landslides and flooding.

In Nashik, another proposed smart city, a citizens' campaign successfully targeted the de-concretization of the Godavari riverbed in the central temple complex area (Nitnaware 2020). This was achieved after a long-drawn legal battle involving a PIL in the Bombay High Court. While the campaign has been widely celebrated as a move towards restoring the river to its 'natural' state, its success has eclipsed the many other concerns of subaltern groups living and working along the river. The main focus of this campaign was reinstating the natural springs in the riverbed that had religious and ritual significance for Hindu devotees. Issues of river pollution, deposition of industrial effluents, and recruitment of manual labour for river cleaning became secondary to the primary concern of religious and aesthetic purity. De-concretizing the riverbed ultimately proved beneficial for controlling seasonal flooding, but the lives of the Dalit and Adivasi workers recruited for manually cleaning the polluted river remained unchanged.[2]

This issue of environmental and climate justice movements in India glossing over questions of caste or remaining indifferent to caste injustice has not gone unnoticed. Scholars and activists have noted that many of the postcolonial environmental movements against resource extraction and displacement have been led by women and Adivasis. However, caste subalterns (especially Dalits) remain under-represented in environmental politics (Sharma 2012).

This absence of recognition takes many forms. The erasure of Dalits from environmental politics has been described as 'eco-casteism' (Sharma 2017). For instance, anti-caste assertions such as the Mahad Satyagraha of 1927, which claimed Dalits' right to use public tanks and demanded equitable distribution of resources, have not been recognized as environmental struggles. Widening the canon of environmental thought in India and recognizing anti-caste philosophers such as Ambedkar as ecological thinkers will allow for greater recognition of caste in environmental discourse (Kumar 2016).

---

[2] Caste and conservation in the context of the Godavari in Nashik have been discussed in detail in an episode of the *Confronting Caste* podcast (Kings India Institute 2021): https://soundcloud.com/kings-india-institute/pollution-conservation (accessed 4 June 2021).

Conclusion: setting an agenda for caste justice in/as climate justice

This chapter has made four key arguments regarding caste and climate justice in India. First, it demonstrated that caste-based extraction has been foundational to capitalist development since colonial times. The colonial government imposed a new property regime that allowed it to extract revenue from colonies that was then invested in building the core infrastructure of the global capitalist system. These property regimes were created through the displacement and dispossession of caste subalterns, the entrenchment of existing caste hierarchies, and the institutionalized deprivation of Dalit–Adivasi groups. Second, the chapter outlined how the absence of property rights continues to shape caste experiences in contemporary India. Drawing on examples from the urban metropolitan context, it shows how the enforced displaceability of caste subalterns continues to fuel urban planning, even while the project of planning itself is entirely dependent on the labour of displaced migrant workers. This shows the continuing role of caste-based extraction in sustaining growth and development in postcolonial India. Third, through the example of urban sanitation work, the chapter has highlighted how the integration of digital technology in urban development has been a bane rather than boon for the predominantly Dalit sanitation workers. The opportunity for deploying technology to eliminate caste-based extraction has been ignored in favour of creating tools of surveillance that further caste-based exploitation. Finally, the chapter has shown the wide chasm between anti-caste politics and urban environmental movements. While anti-caste struggles across the country have made visible the close connection between ecological crises and caste injustice, urban environmentalists have largely failed to account for this in their agendas or even to recognize caste as an important factor in thinking about the environment.

Based on these arguments, I propose that caste injustice has been a constitutive force in bringing about the climate crisis we are faced with today. Moreover, in contemporary attempts to combat climate injustice at the national and international levels, the issue of caste has remained marginalized. This has rendered attempts at climate justice partial at best and self-defeating at worst. A comprehensive and meaningful approach towards tackling the climate crisis would need to fundamentally reimagine climate justice as caste justice. To be able to do this, our scholarly and activist efforts must: (*a*) recognize caste as a key factor that governs human–environment interactions, (*b*) explore and understand the deep histories of caste-based extractions, and (*c*) adopt an anti-caste approach towards mapping the distributional and procedural features of India's nascent climate movements.

References

Adger, Neil, and Helen Adams. 2013. 'Migration as an Adaptation Strategy to Environmental Change'. In *World Social Science Report 2013: Changing Global Environments*. Pasis: OECD Publishing and Unesco Publishing, 261–264. https://read.oecd-ilibrary.org/social-issues-migration-health/world-social-science-report-2013/migration-as-an-adaptation-strategy-to-environmental-change_9789264203419-40-en (accessed 4 June 2021).

Ambedkar, B. R. 2004 [1944]. *The Annihilation of Caste*. New York: Columbia Centre for New Media Teaching and Learning, Columbia University. https://ccnmtl.columbia.edu/projects/mmt/ambedkar/web/readings/aoc_print_2004.pdf (accessed 4 June 2021).

Baviskar, A. 2020. 'Cows, Cars and Cycle-rickshaws: Bourgeois Environmentalists and the Battle for Delhi's Streets'. In *Elite and Everyman*, edited by Amita Baviskar and Raka Ray. New Delhi:Routledge India, 391–418.

Bhan, G. 2016. *In the Public's Interest: Evictions, Citizenship, and Inequality in Contemporary Delhi*, vol. 30. Athens, GA: University of Georgia Press.

———.2017. 'Where Is Citizenship? Thoughts from the Basti'. *Comparative Studies of South Asia, Africa and the Middle East* 37 (3): 463–469.

Bhandar, B. 2019. *Colonial Lives of Property: Law, land, and Racial Regimes of Ownership*. London: Duke University Press.

Bhambra, G. K. 2020. 'Colonial Global Economy: Towards a Theoretical Reorientation of Political Economy'. *Review of International Political Economy* 28 (2): 1–16.

Bhattacharya, N. 2019. *The Great Agrarian Conquest: The Colonial Reshaping of a Rural World*. Albany, MA: SUNY Press.

Björkman, L. 2015. *Pipe Politics, Contested Waters: Embedded Infrastructures of Millennial Mumbai*. Durham, NC: Duke University Press.

Breman, J. 1996. *Footloose Labour: Working in India's Informal Economy*, vol. 2. Cambridge, UK: Cambridge University Press.

Chu, E., and K. Michael. 2019. 'Recognition in Urban Climate Justice: Marginality and Exclusion of Migrants in Indian Cities'. *Environment and Urbanization* 31 (1): 139–156.

Dalit Camera. 2016. 'Jignesh Mevani on #ChaloUna and Beyond'. *Dalit Camera*, 23 August. https://www.dalitcamera.com/jignesh-mevani-chalouna-beyond/ (accessed 4 June 2021).

Das, S. 2018. 'Govt Must Get Its Hands Dirty to Rescue Manual Scavengers'. *Down To Earth*, 1 October. https://www.downtoearth.org.in/news/waste/govt-must-get-its-hands-dirty-to-rescue-manual-scavengers-61756 (accessed 4 June 2021).

Datta, A. 2019. 'Shimla: The Slow Contoured Life of an Imagined Smart City'. Learning from Small Cities. https://www.smartsmallcity.com/blog (accessed 4 June 2021).

Desai, D. 2020. '282 Deaths in Last 4 Years: How Swachh Bharat Mission Failed India's Manual Scavengers'. *The Print*, 27 January, https://theprint.in/india/282-deaths-in-last-4-years-how-swachh-bharat-mission-failed-indias-manual-scavengers/354116/ (accessed 4 June 2021).

Deshingkar, P., and S. Akter. 2009. 'Migration and Human Development in India'. Human development research paper 2009/13, United Nations Development Funds. https://mpra.ub.uni-muenchen.de/19193/ (accessed 4 June 2021).

Ghertner, D. A. 2011. 'Green Evictions: Environmental Discourses of a "Slum-free" Delhi'. In *Global Political Ecology*, edited by Richard Peet, Paul Robbins, and Michael Watts. London: Routledge, 145–165.

Guha, R. 1982. *A Rule of Property for Bengal: An Essay on the Idea of Permanent Settlement.* Hyderabad: Orient Longman.

Harriss-White, B. 2003. *India Working: Essays on Society and Economy*, No. 8. Cambridge, UK: Cambridge University Press.

Hindustan Times. 2018. 'The State Should Come to the Rescue of the Landless Dalit Farmer in India'. Editorial, 5 March. https://www.hindustantimes.com/editorials/the-state-should-come-to-the-rescue-of-the-landless-dalit-farmer-in-india/story-GuvPIRYjZ7J2DG6E6Z02PK.html (accessed 4 June 2021).

Kashwan, P. 2017. *Democracy in the Woods: Environmental Conservation and Social Justice in India, Tanzania, and Mexico.* New York: Oxford University Press.

Khaira, R. 2020. 'Surveillance Slavery: Swachh Bharat Tags Sanitation Workers to Live-Track Their Every Move'. *Huffpost*, 18 February. https://www.huffpost.com/archive/in/entry/swacch-bharat-tags-sanitation-workers-to-live-track-their-every-move_in_5e4c98a9c5b6b0f6bff11f9b (accessed 4 June 2021).

Khan, S., P. Taraporevala, and M. H. Zérah. 2018. 'Indian Smart Cities: Shared Challenges, Variegated Trajectories'. *Flux* 114 (4): 86–99.

Kumar, V. M. R. 2016. 'History of Indian Environmental Movement: A Study of Dr B.R. Ambedkar from the Perspective of Access to Water'. *Contemporary Voice of Dalit* 8 (2): 239–245. https://doi.org/10.1177/2455328X16662371.

Li, T. M. 2017. 'The Price of Un/Freedom: Indonesia's Colonial and Contemporary Plantation Labor Regimes'. *Comparative Studies in Society and History* 59 (2): 245–276.

Ministry of Urban Development. 2015. 'Smart Cities Mission Statement & Guidelines'. https://smartnet.niua.org/sites/default/files/resources/smartcityguidelines.pdf (accessed 4 June 2021).

Monahan, Torin. 2018. 'The Image of the Smart City: Surveillance Protocols and Social Inequality'. In *Handbook of Cultural Security.* Cheltenham, UK: Edward Elgar Publishing.

Moneycontrol News. 2020. 'Using GPS-enabled Watches to Track Sanitation Workers: All about Nagpur Municipal Corporation's Surveillance System'. 28 July. https://www.moneycontrol.com/news/india/using-gps-enabled-watches-to-track-sanitation-workers-all-about-nagpur-municipal-corporations-surveillance-system-5609711.html (accessed 4 June 2021).

Nitnaware, H. 2020. 'Godavari Was Being Buried under Concrete: Until This Man Fought a 5-Year Battle'. The Better India, 16 September, https://www.thebetterindia.com/238029/devang-jani-nashik-remove-concrete-court-case-river-godavari-riverbed-environment-activist-inspiring-him16/ (accessed 4 June 2021).

Patnaik, U. 2017. 'Revisiting the "Drain", or Transfer from India to Britain in the Context of Global Diffusion of Capitalism'. In *Agrarian and Other Histories: Essays for Binay Bhushan Chaudhuri*, edited by S. Chakrabarti and U. Patnaik. New Delhi: Tulika Books.

Pawar, T. 2019. 'Farmers Agree to Give Land for Greenfield Project'. *Times of India*, Nashik edition, 21 August. https://timesofindia.indiatimes.com/city/nashik/farmers-agree-to-give-land-for-greenfield-project/articleshow/70760895.cms (accessed 4 June 2021).

Ramesh, A. and V. Raveendranathan. 2020. 'Infrastructure and Public Works in Colonial India: Towards a Conceptual History'. *History Compass* 18 (6). https://doi.org/10.1111/hic3.12614.

Ranjan, R. and P. Kashwan. 2021. 'Echoes from the Woods: At the Crossroads of Forest Struggles and Human Rights in Postcolonial India'. *International Journal of Human Rights* 25 (7): 1089–1093. DOI: 10.1080/13642987.2021.1884853.

Ray, R. 1974. 'Land Transfer and Social Change under the Permanent Settlement: A Study of Two Localities'. *Indian Economic and Social History Review* 11 (1): 1–45.

Ranganathan, M. 2018. 'Rule by Difference: Empire, Liberalism, and the Legacies of Urban "Improvement"'. *Environment and Planning A: Economy and Space* 50 (7): 1386–1406.

Robinson, C. J., and H. L. T. Quan 2019. *On Racial Capitalism, Black Internationalism, and Cultures of Resistance*. London: Pluto Press.

Roy, T. 2013. 'Rethinking the Origins of British India: State Formation and Military-Fiscal Undertakings in an Eighteenth-Century World Region'. *Modern Asian Studies* 47 (4): 1125–1115. https://doi.org/10.1017/S0026749X11000825.

Safai Karmachari Andolan. n.d. 'Who Is Involved in Manual Scavenging?' https://www.safaikarmachariandolan.org/crisis (accessed 4 June 2021).

Sharma, A. 2019. 'Why Shimla Smart City Projects Worth Rs 2905 Crore Look Destined to Fail'. *Citizen Matters*, 1 August. https://citizenmatters.in/shimla-smart-city-projects-progress-and-obstacles-13445 (accessed 4 June 2021).

Sharma, M. 2012. 'Dalits and Indian Environmental Politics'. *Economic and Political Weekly* 47 (23): 46–52.

———.2017. *Caste and Nature: Dalits and Indian Environmental Policies*. New Delhi: Oxford University Press.

Shankar, S. and K. Swaroop. 2021. 'Manual Scavenging in India: The Banality of An Everyday Crime'. *CASTE/A Global Journal on Social Exclusion* 2 (1): 67–76.

Singh, S. and A. Thorat. 2014: 'Caste Restricted Occupational Mobility: Have Reforms Compelled Employers to Be Caste Blind?' *Himalayan Journal of Contemporary Research* 3 (2): 2319–3174.

Sircar, S. 2018. '"You Can Call It a Mufassil Town, but Nothing Less": Worlding the New Census Towns of India'. *Geoforum* 91: 216–226.

Smart Cities Council. 2020a. 'Greenfield Township on the Anvil in Nashik Smart City', 23 August. https://india.smartcitiescouncil.com/article/greenfield-township-anvil-nashik-smart-city (accessed 4 June 2021).

———. 2020b. 'New Parking, Parks and Wider Roads on the Cards for Shimla Smart City', 11 September. https://india.smartcitiescouncil.com/article/new-parking-parks-wider-roads-cards-shimla-smart-city (accessed 4 June 2021).

Swachh Bharat Mission – Grameen. n.d. 'Vision and Objectives', https://swachhbharatmission.gov.in/SBMCMS/about-us.htm (accessed 4 June 2021).

Taraporevala, P. 2018. 'Demystifying the Indian Smart City: An Empirical Reading of the Smart Cities Mission'. Working paper, Centre for Policy Research, New Delhi. https://www.cprindia.org/research/papers/demystifying-indian-smart-city-empirical-reading-smart-cities-mission (accessed 4 June 2021).

Reimagining Climate Justice as Caste Justice

Thekaekara, M. M. 2016. 'The Dalit Fightback at Una Is India's Rosa Parks Moment'. *The Wire*, 13 August. https://thewire.in/rights/the-dalit-fightback-at-una-is-indias-rosa-parks-moment (accessed 4 June 2021).

Thorat, Sukhedeo and Katherine S. Newman, eds. 2010. *Economic Discrimination in India*. New Delhi: Oxford University Press.

Vijayabaskar, M. and A. Wyatt. 2013. 'Economic Change, Politics and Caste: The Case of the Kongu Nadu Munnetra Kazhagam'. *Economic and Political Weekly* 48 (48): 103–111.

Water Mother by Anupriya

CHAPTER 9

Intersectional Water Justice in India

At the Confluence of Gender, Caste, and Climate Change

Vaishnavi Behl and Prakash Kashwan

Introduction

India's water crisis has been widely covered in the international and national press. In the summer of 2019, the *New York Times* published a series of reports and features on the prospects of Chennai and other large Indian cities running out of water (Subramanian 2019). Data on the absolute scarcity of water, sometimes illustrated using dramatic satellite imageries of water bodies, often dominate discussions on India's water crisis (Sengupta 2019). Recall, for example, the 2019 report about 21 Indian cities running out of groundwater by 2020 (ANI 2019a). We are now well past the dreaded summer of 2020 but there has been no follow-up reportage. The argument that India's water crisis lingers because its effects are experienced unequally along multiple dimensions – caste, class, and gender – is hardly a controversial one for scholars. However, there has been very little discussion in both the international and domestic press of the gross inequalities in access to water. The *New York Times* report mentions the poor, while another describes how women sacrificed daily showers so that office-going male members of the family could afford the luxury instead (Denton and Sengupta 2019). An overwhelming focus on water scarcity instead of water inequalities, we argue, is one of the major causes for the perpetuation of India's water crisis. In this chapter, we seek to examine how the intersection of social inequalities and climate change contributes to water injustice.

In India, access to water is determined by a complex entwining of caste, class, and gender identities that work to perpetuate structural inequalities. While geography and the quality of physical infrastructure greatly influence the extent of water insecurity, they are not entirely responsible for it. In some parts of the country, safe drinking water is inaccessible, causing widespread suffering, illness, and disease. In other regions, cheap and state-subsidized access to water is taken for granted and easily abused (Fatah 2013). The army cantonment and government districts in Delhi receive 375 litres of water per capita per day. On the other hand, South Delhi's Sangam Vihar, an area with a large number of 'unauthorized colonies' and home to many lower-income religious minorities, receives a meagre 40 litres of municipal water per person. In the Bhalswa dairy district, just 30 kilometres from Sangam Vihar, water from community water sources and hand pumps is highly toxic and polluted, so residents are compelled to queue up and collect their quota of drinking water from the government tanker that supplies water to the area once a day (Reuters 2019). In Pune, a bustling metropolis in Maharashtra, each person uses 352 litres of water per day, while the residents of Latur, another district in the same state, are forced to make do with a mere 60 litres per person per day (Waghmare 2016).

Evidently, social and cultural norms and regional inequalities create formidable barriers against equitable access to water. As the impacts of climate change become more prominent, people will experience heightened vulnerability along similar axes. In this chapter, we intervene in an ongoing, vibrant debate on the complexities of water access and water justice (Joshi 2015). We use a framework of intersectional water justice to address some of the analytical gaps that scholars of water justice in India have very insightfully highlighted.

First, we choose to focus on access to clean and safe drinking water, even though we acknowledge that water and sanitation must go together. Water and sanitation are also conceptualized together in the United Nation's Sustainable Development Goals for 2030 and in many debates on human rights to water and sanitation. However, for the sake of analytical clarity, we focus on what we consider to be one of the most urgent social and environmental concerns of the twenty-first century – the lack of access to safe drinking water. Second, while recognizing the importance of policies and governance in determining the outcomes of water allocation and access, we approach the question of water justice from the bottom up to map how multiple social inequalities constrain household access to water (Roth et al. 2018). The goal is to investigate how social and economic inequalities shape access to and the benefits of policies and programmes related to the provision of drinking water, including those led by non-governmental organizations.

In the next section we discuss inequalities in access to water in India. The goal is to highlight the gravity of the situation and communicate the importance of applying an intersectional approach to water justice. This is followed in the third section by a brief discussion of the concept of intersectionality; we apply it to analyse how class, caste, and gender inequalities produce unique experiences of water-based vulnerabilities. We illustrate these arguments in the fourth section by developing grounded analyses of access to drinking water in two different regions First, we examine how, within the context of 'tubewell capitalism' gender and class inequalities intersect to disproportionately burden poor women from the so-called lower castes in north Gujarat (Dubash 2002). These women are forced to walk long distances to fetch drinking water for the household. Second, we analyse the dynamics around access to freshwater sources in the state of Uttarakhand.

In these two case analyses, we emphasize how socio economic and gender inequalities create formidable barriers that limit access to drinking water for marginalized groups, with or without climate change. We do this to avoid the 'climate bandwagon effect', which translates, in many cases, to a reductive framing of important questions as simplistic arguments about the effects of climate change (Jinnah 2011). Having demonstrated the presence of deep-seated social inequalities that shape access to drinking water, we also demonstrate how these social barriers undermine community-based climate adaptation and resilience programmes. In the concluding section of this chapter, we summarize our key arguments and reflect on the analytical benefits and challenges of employing an intersectional approach to water justice.

Background: water scarcity meets social inequalities
As water scarcity has become a lived reality for millions, national and international institutions have amplified the need for water-specific legal instruments or legislations. In 2010, the United Nations General Assembly formally recognized the Human Right to Water and Sanitation (HRtWS). In practice, the realization of HRtWS would mean the provision of safe water in sufficient quantities (that is, 50–100 litres of water per person per day) at affordable costs (such that water costs do not exceed 3 per cent of the household income).

The right to water is implicit in Article 21 of the Indian Constitution, which guarantees the right to life. However, the debates on the topic are entangled in the questions of whether to conceptualize water as a private commodity or part of the commons (Moench 1998). A National Commission that was mandated to review the Constitution, recommended in 2002 that a new Article 30D be inserted to recognize that '*Every person shall have the right—(a) to safe drinking water ...*' but this recommendation is yet to be implemented (Upadhyay 2011, 57, italics added for emphasis). Despite international impetus and national debates, India continues to face an increasingly serious water crisis. According to one estimate, in 2015, only 46 per cent of Indian households had access to piped water supply. WaterAid UK estimates suggest that around 7.6 crore people in India do not have assured access to safe, drinking water, making India home to the largest population without access to safe drinking water (Burgess 2016). India ranked 120 out of 122 countries in the Safe Water Index released in December 2019 (*Down to Earth* 2018; *FnBnews.com* 2019). According to the analysis presented in the *Aqueduct Water Risk Atlas* prepared by Washington DC-based World Resources Institute, India is the 13th most water-stressed country among the 150 countries they analysed (Dormido 2019).

These large gaps in access to safe drinking water, despite vast reservoirs of traditional knowledge on the management of water resources and decades of 'double-digit' economic growth in India, suggest that the root cause of the problem is neither lack of knowledge nor scarcity of resources. The alarming state of the drinking water crisis cannot be blamed entirely on water scarcity, as huge quantities of water are also wasted routinely. Rather, the crisis is the result of the deplorable state of policy-making processes and inequalities related to the distribution and allocation of safe drinking water. As such, the challenge of achieving water justice in India is not about framing access to water as part of the right to life, but that this right must also be universally implemented (Cullet 2013). Thus, the real measure of such interventions lies in how effectively such legal and normative discourses are translated into substantive outcomes in practice. While we recognize the ways in which climate change exacerbates water deprivation, a deeper understanding of the social roots of water inequalities is also crucial in the search for enduring solutions to India's water woes. In this spirit, and to counter climate bandwagoning of water inequality, it is helpful to situate the present analysis within a broader historical context.

The Mahad Satyagraha of March 1927, led by Dr B. R. Ambedkar, in conjunction with the conference organized by the Bashikit Hitakarini Sabha, offers an instructive point of departure and a source of inspiration in this context. Through the Satyagraha, Dalits sought to claim their right to water enshrined in a resolution adopted by the Bombay Legislative Council in August 1923, which stated that 'the Untouchable classes be allowed to use all public watering places, wells, Dharmashalas which are built and maintained out of public funds, or are administered by bodies appointed by Government' (Moon et al. 2014). Despite such a resolution, the Dalits of Mahad continued to be denied water from the public tank. To fight this, on 20 March 1927,

Ambedkar led a march of about 2,500 participants of the Dalit conference to a public tank in the centre of Mahad for a symbolic drink.

Even though it was a peaceful protest, upper-caste men from the village mounted a violent attack against the caravan led by Ambedkar (Moon et al. 2014). Keeping with segregationist casteist norms, upper-caste Hindus performed purification ceremonies to cleanse the well of the 'sins' of defilement caused by the protestors (Narake et al. 2003, 12). Later that year, when Ambedkar and his followers decided to hold a second Satyagraha, caste Hindus successfully secured a judicial injunction against any attempt to draw water from the well. Unwilling to break the law, on the night of 25 December, Ambedkar and his followers returned to Mahad and publicly burnt copies of the *Manusmriti*, the book of conservative Hindu laws that equates women and untouchables to cattle. This second Satyagraha also saw an exponential increase in the participation of women. Ambedkar inspired women to challenge the practice of untouchability, emphasizing their responsibility to the movement (Deepa 2017). Ambedkar's dual approach to this struggle – that is, his focus on access to water, which is indispensable to living a dignified life, along with the broader agenda of restructuring Indian society – suggests why this fundamental right continues to be elusive in the Indian social hierarchy.

The deep inequalities within Indian society and culture produce unequal access to water along the axes of caste, class, and gender. In the face of increasingly precarious access to water, girls and women spend an extraordinary amount of time collecting potable water. In areas without accessible freshwater sources, young girls from poor families walk long distances to fetch water, making them vulnerable to various types of harassment and violence. Some are forced to sacrifice their education to carry out these domestic chores, as the high school dropout rate for girls in India indicates (Seymour 2020). Not all women experience this exclusion similarly, as norms related to untouchability and segregation add to the difficulties of Dalit women. As we discuss later in this chapter, menstruating lower-caste women find themselves without access to any water when they need it the most. A rigorous understanding of these disadvantages, therefore, requires an intersectional understanding of water justice. The next section offers a brief introduction to the framework of intersectionality.

Water rights, social inequalities, and intersectionality: a conceptual discussion

The United Nations General Assembly formally recognizes the HRtWS as part of international legal norms, thereby creating an impetus for governments to integrate it into national policies and plans. The international and national legal recognition of HRtWS makes it an internationally accepted normative framework that activists and policymakers can use as a reference (Zwarteveen and Boelens 2014). However, in most cases, such international instruments do not account for the countervailing force of neoliberal reforms, which undermine state interventions in the provision of basic civic amenities (Kashwan, Maclean, and García-López 2019). They also, of course, cannot account for the social and political barriers that prevent the realization of justice and human rights in practice.

Legal precedents from various Indian courts recognize the right to water as indispensable to the right to life enshrined in the Indian Constitution, but the Indian state is yet to come up with legally enforceable provisions for the execution of these rights (Upadhyay 2011). While the legal provisions are necessary, they are insufficient for the realization of universal, safe, and affordable access to drinking water. An important part of this challenge is the social reality that hampers universal access to safe drinking water. The neoliberal policies and programmes of the last quarter-century, as well as the ostensibly welfare-oriented policies in the past, relied on 'tokenistic ... and apolitical' references to gender and caste concerns, which have only reinforced pre-existing inequalities (Joshi 2011, 56). Moreover, the limitations of the legal process also apply to policies, which are necessary but rarely sufficient to bring about transformative changes. For example, the oft-cited 'traditional' water harvesting systems, such as Dharas, exclude Dalits (Krishnaraj 2011). Some of the best non-governmental organizations (NGOs) and social movements struggle to reach the most marginalized because of the socio-economic realities that keep them from participating in community-based institutions or rights-based social mobilization (Kashwan and Lobo 2014).

Effective legal, policy, and programme development require a deeper understanding of the workings of social inequalities linked to caste, class, religion, and gender. These social 'structures' that shape access to safe drinking water are entrenched in the everyday life of the marginalized via tacit codes of conduct that maintain and perpetuate caste-based segregation, practices of untouchability, and gender-based inequalities in both rural and urban areas (Mehta 2016). Some types of policy work seek to sidestep these problematic aspects of access to water and other natural resources by framing water scarcity as a 'natural' and inevitable result of climate change, without considering the effects of severe imbalances in social, economic, and political power (Zwarteveen and Boelens 2014). Such efforts at naturalizing social inequalities are bound to end in disappointment, especially if the goal is to promote just and enduring responses to climate change. To shine a light on the folly of depoliticizing environment and development interventions and legislation, we adopt a framework of intersectional water justice and illustrate its application in researching the water injustices faced by the members of socially, culturally, and politically marginalized groups.

Originally conceptualized by Kimberlé Crenshaw as a mechanism to analyse the marginality of African-American women, intersectionality emphasizes the cross-cutting nature of socio-cultural and economic inequalities (Crenshaw 1989). While it is essential to investigate how the disadvantages of gender, caste, and class limit access to drinking water, a failure to investigate their intersections produces an incomplete, even distorted view of reality (Kaijser and Kronsell 2014).

In its simplest form, intersectionality has been used to highlight the *convergence* of multiple forms of oppression – for example, the double disadvantages experienced by lower-caste women. However, intersectionality also helps us focus on 'privilege and demonstrates that intersectionality and interlocking oppressions are time and context contingent' (Hulko 2009, 44). In the Indian context, intersectionality points to how injustice is experienced differently by upper-caste women compared to lower-caste woman; or by relatively well-off lower-caste women as compared to poor lower-caste men. In other words, due to the complex intersections of the multiple layers of identity, the same person may be either privileged or marginalized in different contexts or at different times (Thompson 2016).

Interdisciplinary research on environment and development contribute important insights that are especially relevant for studies of intersectional water justice. One, environmental social scientists show how embodied experiences of social difference – such as gender, caste, and religion – are produced and expressed through everyday material realities, including those linked to the use of natural resources (Nightingale 2011, 154). Two, in many cases, outcomes of contested claims to material resources may reveal additional layers of (dis)advantage and alter the socio-economic status of claimants. Three, by making visible the broader political economy of resource control, the environmental social sciences draw attention to the structural politics of resource poverty and deprivation that a disproportionate focus on social difference may disguise (Rao, Min, and Mastrucci 2019). Each of these insights is also relevant to the contestation over societal responses to climate change. The possibility that those with certain types of socioeconomic privileges may benefit from the disruptions of a climate-changed world adds another layer to an intersectional analysis of climate justice. The use of intersectionality as an analytical tool highlights the importance of carefully analysing the differentiated and cross-cutting nature of inequalities and privileges. We reflect on some of these possibilities in analysing the intersectionality of water justice.

Climate stresses meet caste, class, and gender inequalities: two cases

Access to groundwater in rural Gujarat

Groundwater is a major source of drinking water in India. This is especially true of the arid and semi-arid regions of North Gujarat, which stand on top of rechargeable aquifers. The region relies on groundwater to meet its water requirements amidst pervasive scarcity owing to unpredictable and scanty rainfall. Additionally, increasing salinity of coastal aquifers presents new threats of water stress. Water scarcity is also strongly linked to the increased use of tubewells, which has led to nearly unchecked extraction of groundwater for irrigation (Bhatia 1992).

Landowners hold *de facto* property rights over groundwater. However, the interconnectedness of underground aquifers means that groundwater reservoirs are effectively open-access resources. The extraction of groundwater is constrained only by the availability of means of extraction, such as tubewells, electricity, or other sources of energy (Moench 1992). This means that access to groundwater is directly tied to economic status and inequalities. Casteist social structures that restrict property inheritance and ownership of land among the lower castes also explain disparities in access to water. According to one study in the Kutch district, there is one tubewell for every 1.4 Patidar households, but the ratio falls to one well per 30 families among the lower-caste group, the Vankars (Bhatia 1992). Moreover, these inequalities have worsened with the introduction of modern water extraction technologies, such that all tubewells in the village Bhatia studied are owned by the Patidar caste, who constitute a mere 23 per cent of the village population.

By heavily investing in tubewells, upper castes in Gujarat and elsewhere free-ride on underground aquifers that should ideally be managed and regulated as commons. Exploitation of groundwater has so severely depleted water tables that it threatens a complete collapse of groundwater resources. Moreover, because the intersection of caste- and class-related inequalities shapes inequalities in the distribution of modern technologies, the collapse of underground aquifers is rooted in – and further reinforces – socio-economic inequalities.

The environmental collapse caused by socially and economically powerful groups directly contributes to poor and lower-caste farmers' dispossession of their rightful share of groundwater, including safe drinking water. For example, the residents of Mathnaa village in Sabarkantha district in north-east Gujarat have relied historically on shallow dug wells for their supply of irrigation and potable water. However, ever since the upper-caste families started using tubewells to extract water faster from the aquifer, shallow wells no longer supply sufficient water. The resulting exacerbation of water inequalities led to a concentration of power in the upper-caste households, who are also known as the 'Water-Lords' (Naz 2015, 89). Poor farmers, who are unable to invest in costly tubewells, became dependent on the upper castes for fulfilling their water requirements, which pushed them further down the spiral of social and economic subordination.

Tubewell-assisted exploitation of aquifers is not necessarily an example of improved overall social welfare at the cost of environmental sustainability – in most cases, it exacerbates social inequalities while also undermining the sustainability of groundwater in these arid and semi-arid regions. However, this does not mean that communities can switch back to some form of traditional water harvesting system. In many cases, better regulated and socially equitable management of tubewell-based groundwater extraction seems to be the only option. One such reform advocated for by small and marginal farmers in water-scarce North Gujarat is to make tubewells a public good. They propose that the pipe that runs through the ground should be considered public property, equally shared by all members of the community, while individual families or smaller groups of families may invest in motors or pumps to extract water at a small cost (Bhatia 1992). This would reduce the monopoly of land-owning farmers over water resources and might also help regulate the amount of water that upper-caste families extract.

In the arid and semi-arid states of Rajasthan and Gujarat, women spend an average of three hours collecting drinking water from wells – which, along with public taps, form one of the two major sources of drinking water in the region. This is a major source of drudgery in their daily lives (Varua et al. 2018). The drinking water supply in arid and semi-arid regions is often controlled by state agencies charged with the task of supplying water in dry months. For example, the Gujarat Water Supply and Sewerage Board (GWSSB) is a statutory body in charge of managing and regulating water supply in the state. However, like many other state institutions, the GWSSB's performance is often erratic. The water that they supply is also of inferior quality, supplied on alternate days, and amounts to an average of five litres per person per day (Agarwal 2019). This falls short of the minimum water requirements of 20 litres per person per day, as laid out by the United Nations (Watkins 2006). However, such scarcity of water is not exceptional.

In villages throughout the northern and western parts of the state, wells often run out of water by early March. However, GWSSB operates water tankers from late April until early May. In the meantime, lower-caste villagers must rely on the wells and tubewells owned by the upper castes or travel miles to get water from faraway public tanks (Prakash and Sama 2006; Kulkarni et al. 2020). In Merka, a village in Kutch with no reliable source of safe drinking water, GWSSB tankers are supposed to supply water three times a day. However, the tankers show up erratically and sometimes not at all; when they do appear, they supply poor-quality water that is too saline to drink. In most cases, villagers use the water only for washing, cleaning, and other domestic tasks, but rely on village wells for their supply of drinking water (Mehta 1997). Even in villages where the state has installed public standposts to provide sufficient quantities of potable and drinking water, caste-based discrimination limits lower castes' access to water, sometimes leading to violent attacks on Dalits (Mehta 1997). A study of 1,589 villages in the state of Gujarat showed that Dalits are denied access to public drinking water infrastructure in 29 per cent of villages, and Dalit settlements have no public taps or wells in 71 per cent of the villages (Armstrong and Davenport 2010).

The GWSSB's unreliable operation of water tankers exposes Dalit women to increased vulnerability and harassment, as it renders them entirely dependent on the upper castes for employment and drinking water supply (Prakash and Sama 2006; Armstrong and Davenport 2010). Dalit and other lower-caste women are therefore subject to intersectional injustices linked to the concentration of multiple vulnerabilities arising out of their caste and gender identities. They are relegated to a life of extreme marginality and experience continuous physical and sexual threats by upper-caste men, and verbal abuse and taunts by upper-caste women (Dhar 2017). Patriarchal norms that exist within the community are often replicated within the household, making Dalit women highly vulnerable to intra-household violence. While instances of violence against Dalit women are most often observed in public spaces, like open fields and streets, the second-most common space for violence is the household. Dalit women's efforts to challenge patriarchal authority often result in domestic violence, which points to Dalit men's acceptance and perpetuation of gender-based inequalities (Irudayam, Mangubhai, and Lee 2006).

The cultural politics of access to and control over water infrastructure in Uttarakhand

In several regions across India, especially in the hills, perennial streams are considered sacred. The Himalayan state of Uttarakhand is endowed with major riverine networks that feed several traditional water systems to bring water to mountain villages (Acharya 2011). Highly localized ecological knowledge systems, developed over several hundreds of years, enable indigenous communities to cope with drastic environmental changes and hold the key to culturally appropriate climate adaptation strategies. However, patriarchal and casteist structures define customary sacred norms around these local water systems. The purity–pollution divide, which is central to Hindu rituals, means that the access of Dalits and menstruating women remains fragile and highly restricted, thus severely damaging lower-caste women's ability to withstand the impacts of climate change.

Dharas, traditional water systems where subterranean or spring water is directed to the surface through well-adorned and specially carved outlets, are a crucial source of domestic water supply in the Garhwal region. The norms around the use of water from a *dhara* are heavily embedded in socio-cultural structures and religious symbolism. A *dhara* is marked by a small shrine, dedicated to a deity, either inside or close to the spring. Small temples are often constructed close to *dharas*, and in the absence of a temple, the *dhara* itself serves as a place of worship. A selective understanding of religious norms associated with traditional water harvesting structures can produce a misleading picture of the lived experiences of different groups within a community. In Chunni village, upper-caste Khanka Kshatriya women attributed the abundant availability of water to the blessings of Jal Devi and claimed that there was more than enough water to use in daily rituals (Joshi 2011). On the other hand, Dalit women in the same village face acute water scarcity – they are subjected to water apartheid, as they are assigned separate *naulas* (a water harvesting structure that taps into subterranean springs) only from which they are allowed to collect water (Joshi 2011). Some of them are compelled to reuse wastewater from cleaning utensils and laundry as feed for farm buffaloes. Young Dalit girls who are sent to collect water from *naulas* are often subject to sexual harassment and violence; there have even been instances where men have raped girls as young as nine years old (*Hindustan Times* 2019).

In Rautgara village in Pithoragarh tehsil in Kumaon division, menstruating girls are forbidden from going to school for at least five days to avoid 'contaminating' a temple that is located on their way to the school (Punetha 2018). Menstruating girls are also forbidden from collecting water and from touching or even coming close to the *dhara* (Acharya 2011). Such norms not only significantly limit their access to water when they need it the most, potentially endangering their health, but also dictate their access to other crucial resources and services. The practice of untouchability also influences how spaces around these traditional water sources are used. In Garhwal, separate *dharas* are assigned to the lower castes depending on their location and water quality (Asthana 2003). In areas with a single *dhara* for all the castes, Dalit women can collect water only after upper-caste women have collected their share (Acharya 2011). The lower castes are often verbally and physically abused for attempting to use upper-caste water sources.

This is not to suggest that no advancements have been made in improving drinking water infrastructure in Garhwal. The introduction of piped water has improved the drinking water supply for Dalit households and reduced the distance that Dalit women need to walk to collect water. However, researchers suspect that these gains are unlikely to be sustainable without the incorporation of traditional water conservation practices and conservation in catchment areas, as there has been a disastrous reduction in the overall water supply (Acharya 2011). More importantly, addressing gender- and caste-based inequalities is crucial for just and equitable access to water. In 2018, a Dalit man was allegedly murdered by two Gujjar men in what started as a squabble over the usage of a water canal in Haridwar (*Scroll.in* 2018). These atrocities reflect the deeper social reality of caste-based discrimination in Uttarakhand, where, according to a recent survey, over 50 per cent of upper-caste individuals admitted practising untouchability, and 68 per cent of the Brahmins in rural Uttarakhand confessed to the practice (Thorat 2020). As expected, Brahmins and other upper-caste men also dominate the state's politics and policy-making processes.

Social inequalities and implications for climate adaptation and resilience

Both Gujarat and Uttaranchal are among the most climate-vulnerable states in India, though for very different reasons. While rising atmospheric temperature is associated with melting glaciers and drying of perennial springs in the Himalayas, it is likely to contribute to increased aridity and scarcity of water resources in the semi-arid environments of Gujarat. Despite these and other differences, the social realities of caste and gender present similar stories of exclusion, oppression, and overburdening of women with households responsibilities, especially Dalit women.

Scholarly and activist arguments in favour of aggressive and timely climate action are often justified in the name of social justice concerns – such as the argument that climate adaptation and resilience are likely to alleviate the disproportionate burden that climate change imposes on marginalized groups. Yet these arguments are accurate only in a very superficial sense; they are premised on the assumption that addressing climate change will also address its disproportionate social effects without active interventions targeting deeply embedded social inequalities. Second, even when social inequalities are considered, climate policies and programmes cannot be made gender-sensitive simply by 'adding on a concern for women' (Ahmed and Fajber 2009, 33). The intersectional approach helps deepen our understanding of the detrimental effects of the multiple, intersecting disadvantages that individuals face based on their social and cultural positioning. More importantly, as we demonstrate in this section, intersectional inequalities also shape the distribution of costs, burdens, and benefits related to climate adaptation and resilience interventions.

Climate impacts and social responses in Gujarat

The decentralization of decision-making power to community organizations and locally elected governments, along with the intervention of NGOs, are often recommended as the preferred approach to climate resilience and adaptation (DasGupta and Shaw 2014). It is worth considering an example of such a partnership between two very reputed governmental and non-governmental organizations active in Gujarat: (*a*) the Water and Sanitation Management Organisation (WASMO), a governmental agency charged with overseeing community-managed water supply development in rural Gujarat, and (*b*) the Aga Khan Rural Support Programme (India) (AKRSP [I]), an internationally renowned NGO with expertise in provisioning of drinking water in rural areas, which has been working with local communities in the drought prone-district of Surendranagar since 2002.

WASMO and AKRSP joined hands to implement a successful desilting project, which led to improved water availability in Vadali in Surendranagar. As a follow-up project, they proposed the construction of a common drinking-water well close to the check dam to improve water security in the village. However, influential village leaders, each of whom owns large areas of land and has access to sufficient water through their private wells, blocked these efforts. They feared that the newly proposed village well would deplete the aquifer and reduce the amount of water available to them for irrigating their fields. The lower castes, fearing repercussions, were unable to voice their opposition or exercise any bargaining power after the NGOs called off the project. They then chose an alternative site acceptable to the upper-caste households for the construction of a new well, even though it was located outside the area that would have benefitted from the desilting project (Prakash and Sama 2006). We observe similar dynamics in the Earthquake Rehabilitation and Reconstruction (ERR) programme that WASMO implemented in consultation with ASRSP (I) in Surendranagar in 2003. For instance, in Navagam village of Surendranagar district, project staff chose the sites for the construction of water storage tanks as per the wishes of the middle-caste Bharvad community instead of through a scientific analysis of the aquifer (Kulkarni et al. 2020).

Other attempts to decentralize water governance through *pani samitis* (water committees), duly linked to locally elected village councils, have also failed to challenge the dominance of caste- and gender-based community norms (Krishnaraj 2011). In Ghogha and Navagam villages in Kutch and Surendranagar respectively, *pani samitis* have failed to ensure the equitable distribution of water to distant hamlets, which, in most cases, are impoverished Dalit settlements (Kulkarni et al. 2020). Socially and economically powerful households often tamper with water pipes to withdraw additional water than their allocated share more frequently, thereby diminishing water availability for users downstream. The construction of new infrastructure and the constitution of local community organizations, while necessary, are insufficient to alleviate the vulnerability of the most marginalized.

Climate change is likely to stress existing supplies of groundwater, especially for marginalized groups, whose access is already precarious. Gujarat happens to have the longest coastline of all states in India, which makes its groundwater-stressed coastal areas highly prone to the intrusion of salinity. Well water in villages in Jafrabad *taluka* have already become too saline for human consumption, forcing women to walk for over a kilometre to secure their necessary supplies of drinking water (Brahmbhatt and Ved 2019). Many of these communities already recycle household wastewater for multiple domestic chores. However, the problem of salinization of groundwater is not restricted to coastal areas. In Becharaji village of Mehsana district, wells in Dalits neighbourhoods have turned saline. However, Dalits are not allowed to collect water from the community well at the centre of the village. They are expected to collect their drinking water supply from a separate tank constructed some distance from the well, so that the water from the Dalit tank does not mix with the well water; the upper-caste villagers believe that it will 'pollute' the community well (Dhar 2017). Such discriminatory practices are rooted in the local social context, but they are reinforced further as the state fails to invest in basic amenities, leaving marginalized communities to fend for themselves.

The significance of the intersection of longstanding social inequalities with the salinization of groundwater resources goes beyond the state of Gujarat. South Asia is projected to witness a massive, long-term loss of groundwater because of the intrusion of salinity in its coastal areas. Climate models suggest that the region is likely to see an estimated 0.075 per cent loss of fresh groundwater per year in both high emission and low emission scenarios leading up to 2099 (Ranjan, Kazama, and Sawamoto 2006). India's long and densely populated coastline presents formidable challenges for the pursuits of socially just climate adaptation.

Climate impacts and social responses in Uttarakhand

In the age of climate change, the Himalayan region faces the threats of melting glaciers, destabilization and landslides, rapid erosion of terrace farms, and drying of perennial streams and other water sources (Chrisensen 2019). Climate change and frequent forest fires have significantly reduced vegetation and increased surface run-off, reducing the quantity of potable water and transforming once perennial water springs into seasonal sources, with less and less water able to percolate through the terrain every year (Acharya 2011). Scholars are debating whether migration from hilly areas is a feasible climate adaptation strategy (Datta 2019). However, the lens of intersectionality should serve as a reminder that the ability to migrate is not distributed equally among different sections of society. This is especially pertinent in light of the precarity of the urban context for migrant workers, especially Dalit and women workers (Sirimane and Thapliyal 2020). Even if migration to urban centres becomes more precarious, administrative neglect, increasing water scarcity, and irregular drinking water supply has driven large-scale outmigration. This is especially true of the Pauri district of Uttarakhand, where many towns are turning into 'ghost towns' (Singh 2019). In most cases, migration from hills is a male phenomenon, leaving women to take care of agriculture and other household affairs (Tata-Cornell Institute for Agriculture and Nutrition 2019–2020).

A gender-differentiated understanding of migration also often prompts NGOs and policymakers to ensure that women are represented and that they even lead the committees for community-based natural resource management and climate adaptation interventions. However, research shows that women from upper-caste landowning families often occupy these remedial spaces, in which they exercise limited decision-making authority over, say, the management of water resources. However, despite intense engagement and knowledge of water resources, Dalit and other lower-caste women face a variety of barriers that prevent them from experiencing the benefits of these representation processes (Khandekar et al. 2019). As the next example demonstrates, this should not be read as an argument in favour of taking responsibility away from local communities or bringing in NGOs for the arbitration of social conflicts.

Deepa Joshi's research on the World Bank-financed project, Swajal, provides important insights on the stubbornness of caste–gender hierarchies and the failure of NGOs and multilateral agencies in accounting for and addressing them. The Swajal project, led by a prominent NGO, employed a significant number of female employees who occupied lower-level positions in the field. However, the project did not make similar efforts to employ Dalit women or men (Joshi 2011, 61). This was perhaps not a coincidence, considering that the local NGO involved in the project, well-respected for its work on community-based environmental and development work, was founded by an all-Brahmin leadership team and did not have any Dalit staff members (Joshi 2011, 61). With such a complete lack of Dalit staff and lack of women in senior positions, the project also produced socially discriminatory effects.

In the village of Mala, which was also the project's flagship village and the first to complete implementation, the project appropriated the water source of a Dalit woman-headed household. Even more tragically, this Dalit household did not receive any benefits from the project. Therefore, notwithstanding the liberal rhetoric of gender-sensitive project strategies, multilateral projects meant to promote community resilience and adaptation could significantly exacerbate the vulnerabilities of marginalized households.

Conclusion

The international media's focus on 'Day Zero', when large metropolitan cities must shut off their metaphorical taps, has drawn attention to the issue of water stress. However, as we mentioned in the introduction, the media's focus on the water crisis in Chennai, as well as its reportage on the 21 Indian cities that would run out of groundwater by the summer of 2020, was characterized by remarkable blindness to deep inequalities in access to safe drinking water. Amidst a deepening water crisis in the summer of 2019, Dalit households living at the foothills of the sacred Otthakadai Yanamalai were barred from drawing water from a public well. Powerful upper-caste community leaders sought to explain this by arguing that the well is 'sacred' and that 'other people are not allowed to visit the well because they are not clean' (ANI 2019b). However, villagers also revealed an equally important source of upper-caste anxiety when they mentioned that more than 150 people already use the water from the same well for drinking purposes. This was not an isolated case, with at least one other report suggesting that Dalits in more than 100 villages faced similar discrimination when accessing public wells (Pal 2019). Climate change-related stresses will certainly exacerbate the social production of discrimination; so, social factors cannot be left aside for post-facto or marginal considerations in NGO projects and government interventions.

Quite evidently, the long and inspiring history of social struggles exemplified by the historical Mahad Satyagraha should continue to guide our present-day thinking on widespread discrimination in access to water. As is evident from the reports that suggest that such incidents worsen during the summer months, climate change-related stresses on water supply are likely to deepen the discriminations that Dalits, especially Dalit women, experience. While intersectionality has become a popular social science concept somewhat recently, for Ambedkar, caste and gender oppression were both rooted in the deeply hierarchical and exploitative philosophy of orthodox Hinduism. This was evident in the protest that Ambedkar staged as a follow-up to the Mahad Satyagraha, where he performed a symbolic burning of the *Manusmriti*, the main source of the anti-woman and anti-Dalit ideology of orthodox Brahminism (Vajpeyi 2016, 5). Yet popular approaches to discussing gender- or caste-related vulnerabilities in the context of climate change do not sufficiently account for the differentiated and intersectional effects of multiple disadvantages in a person's lived worlds.

A deeper analysis of intersectional vulnerabilities to climate change reveals why it is difficult to address these vulnerabilities by introducing apparently gender-sensitive or pro-poor climate adaptation interventions. As we observed in both the Gujarat and Uttarakhand cases, these interventions are invariably appropriated to serve the goals of locally powerful actors. Yet powerful agencies, including multilateral agencies such as the World Bank, have done little to incorporate a nuanced understanding of these realities into their programmes and projects. Instead, they have settled into pro forma responses in the form of local committees, which are endowed with symbolic authority and functional mandates, as a substitute for structural transformations. Such neglect is especially surprising given that the Bank's in-house research has shown how their 'participatory' projects are often subject to elite capture (Mansuri and Rao 2012). Examples from the water sector include attempts to decentralize water governance through *pani samitis*, which have also failed to challenge the dominance of caste and gender-based community norms (Krishnaraj 2011; Kulkarni 2011).

In this chapter, we have sought to shine light on these complacencies and argue for a sustained focus on social power as perhaps the most important ingredient in climate adaptation via water resource management. Socio-political movements that develop 'alternative cultural, social and political paradigms' are a necessity; yet they form an insufficient prerequisite for transformative change (Kulkarni 2011). Large-scale social transformation requires formidable coalitions between state and non-state actors who adopt strategies of 'aggressive partisanship[s]' in favour of socially and politically marginalized groups (Mehta 1997). Moreover, these partnerships would need to marshal a combination of discursive, material, institutional, and ideational powers and counter-powers to displace the stubborn endurance of the status quo (Kashwan, Maclean, and García-López 2019).

We have also sought to contest apolitical discourses that focus on water scarcity without accounting for the effects of various types of inequalities. By developing an intersectional analysis, we showed how individuals experience water insecurity to different degrees depending on the intersection of various layers of their identity. These layers may take the form of one's caste, class, gender, or religion, with their effects contingent on specific historical and cultural settings. For some, as in the case of Dalit women facing verbal, physical, and sexual abuse at community water sources, societal norms function as deeply entrenched, localized, and, yet, normalized forms of crisis. In a country as demographically and topographically diverse as India, addressing a problem as complex as inequities in access to water requires a multi-pronged approach – one that not only addresses the perpetuation of inequalities linked to neoliberal infrastructural development and privatization but also the inequalities embedded in enduring social structures and the rules, norms, and customs that are often celebrated rather uncritically in climate policy circles.

The analyses that we have presented in this chapter should also serve as a cautionary note for external agencies and experts who look at caste-based inequalities as a result of some kind of social anachronism. Top-down neoliberal interventions not only fail to serve the interests of the marginalized in many cases but they also end up reinforcing these unequal structures. Uncritical promotion of community-based natural resource management, especially through the celebration of multiple award-winning social activists despite their superficial engagement with local communities, has undermined the agenda of socially just water resource management (Kashwan 2006). The non-existent Dalit representation in the World Bank-supported Swajal programme in Uttarakhand, and WASMO's appeasement of upper-caste agendas in Gujarat, and agency officials' unquestioning acceptance of these inequalities, require serious administrative and institutional measures. Officials must be held accountable for failing to respond to the apparently discriminatory workings of participatory processes. It is challenging to find 'fixes' to such an incredibly complex issue so deeply entrenched within Indian society, which cannot be addressed easily, especially when the commitment of state machinery to transformative social change is suspect. Even so, it is evident that lasting impacts can only be made via a combination of political and institutional reforms, along with social interventions that undermine the entrenched inequalities within communities and households. Superficial representation and thoughtless legal provisions must be remedied or abandoned entirely. The pursuit of intersectional water justice will be a long-drawn battle, but it is not one that we can ignore any longer.

References

Acharya A. 2011. 'Managing "Water Traditions" in Uttarakhand, India: Lessons Learned and Steps Towards the Future'. In *Water, Cultural Diversity, and Global Environmental Change*, edited by B. Johnston, L. Hiwasaki, I. Klaver, A. Ramos Castillo, and V. Strang. Dordrecht: Springer, 411–432. https://doi.org/10.1007/978-94-007-1774-9_29.

Agarwal, K. 2019. '"We Fear Drought More Than War," Say Border Villagers in Gujarat'. *The Wire*, 25 March. https://thewire.in/rights/gujarat-border-drought-water-crisis (accessed 21 October 2021).

Ahmed, S. and E. Fajber. 2009. 'Engendering Adaptation to Climate Variability in Gujarat, India'. *Gender and Development* 17 (1): 33–50.

ANI. 2019a. '21 Indian Cities Will Run Out of Groundwater by 2020: Report'. *NDTV*, 20 June. https://www.ndtv.com/india-news/21-indian-cities-will-run-out-of-groundwater-by-2020-report-2056129 (accessed 21 October 2021).

———. 2019b. 'Madurai: Amidst Water Crisis, Dalits Restricted from Using Water from a Public Well'. 29 June. https://www.aninews.in/news/national/general-news/madurai-amidst-water-crisis-dalits-restricted-from-using-water-from-a-public-well20190629133733/ (accessed 21 October 2021).

Armstrong, D. and C. Davenport. 2010 *Understanding Untouchability: A Comprehensive Study of Practices and conditions in 1589 villages*. Ahmedabad: Navsarjan Trust and the Robert F. Kennedy Center for Justice & Human Rights. https://www.researchgate.net/publication/46476924_Understanding_Untouchability_A_Comrehensive_Study_of_Practices_and_conditions_in_1589_villages (accessed 21 October 2021).

Asthana, R. 2003. *Evaluation of Varied Approaches for Enabling Sustainable and Equitable Access to Drinking Water in Uttaranchal*. New Delhi: Development Centre for Alternative Policies.

Bhatia, B. 1992. 'Lush Fields and Parched Throats: The Political Economy of Groundwater in Gujarat'. *Economic and Political Weekly* 27 (51/52): A142–A170.

Brahmbhatt, Alok and Mihir Ved. 2019. 'At This Amreli Village, "Water" Pips "Jingoism" as Poll Issue'. *Ahmedabad Mirror*, 18 April. https://ahmedabadmirror.indiatimes.com/ahmedabad/cover-story/at-this-amreli-village-water-pips-jingoism-as-poll-issue/articleshow/68928865.cms (accessed 21 December 2021).

Burgess, T. 2016. *Water: At What Cost? The State of the World's Water 2016*. WaterAid UK. https://washmatters.wateraid.org/publications/water-at-what-cost-the-state-of-the-of-the-worlds-water-2016 (accessed 27 December 2021).

Chrisensen, J. 2019. 'Climate Change Will Melt Vast Parts of the Himalayas, Study Says'. CNN World, 5 February. https://edition.cnn.com/2019/02/04/world/climate-change-himalayas-melt-study/index.html (accessed 21 December 2021).

Crenshaw, K. 1989. 'Demarginalizing the Intersection of Race and Sex: A Black Feminist Critique of Antidiscrimination Doctrine, Feminist Theory, and Antiracist Politics'. *University of Chicago Legal Forum* 1989 (1): 138–167. http://chicagounbound.uchicago.edu/uclf/vol1989/iss1/8 (accessed 21 December 2021).

Cullet, P. 2013. 'Right to Water in India: Plugging Conceptual and Practical Gaps'. International Journal of Human Rights 17 (1): 56–78.

DasGupta, R., and R. Shaw. 2014. 'Role of NGOs and CBOs in a Decentralized Mangrove Management Regime and Its Implications in Building Coastal Resilience in India'. In *Civil Society Organization and Disaster Risk Reduction. Disaster Risk Reduction (Methods, Approaches and Practices)*, edited by R. Shaw and T. Izumi. Tokyo: Springer, 203–218. https://doi.org/10.1007/978-4-431-54877-5_11.

Datta, A. 2019. 'Migration Triggered by Climate Change Is Nearing a Flashpoint: A Ground Report from Gujarat and Uttarakhand'. *Economic Times*, 28 February. https://economictimes.indiatimes.com/prime/environment/migration-triggered-by-climate-change-is-nearing-a-flashpoint-a-ground-report-from-gujarat-and-uttarakhand/primearticleshow/68193655.cms (accessed 21 December 2021).

Deepa, B. 2017. 'Mahad Satyagraha: Dr. Ambedkar's Speech to Enlighten Dalit Women on Social and Cultural Rights'. *International Journal of Research in Education and Psychology* 3 (1): 20–26. http://ijrep.com/wp-content/uploads/2017/04/DR.-AMBEDKARS-SPEECH-TO-ENLIGHTEN-DALIT-WOMEN.pdf (accessed 21 December 2021).

Denton, B. and S. Sengupta. 2019. 'India's Ominous Future: Too Little Water, or Far Too Much'. *New York Times*, 25 November. https://www.nytimes.com/interactive/2019/11/25/climate/india-monsoon-drought.html (accessed 21 December 2021).

Dhar, D. 2017. 'For Dalits in Rural Gujarat, Untouchability Is Still a Part of Everyday Life'. *The Wire*, 29 August. https://thewire.in/caste/gujarat-dalits-untouchability (accessed 21 December 2021).

Dormido, H. 2019. 'These Countries Are the Most at Risk from a Water Crisis'. *Bloomberg*, 6 August. https://www.bloomberg.com/graphics/2019-countries-facing-water-crisis/ (accessed 21 December 2021).

Down to Earth. 2018. 'See How Indian States Fared on the Water Index'. 15 June. https://www.downtoearth.org.in/news/water/see-how-indian-states-fared-on-the-water-index-60868 (accessed 21 December 2021).

Dubash, N. K. 2002. *Tubewell Capitalism: Groundwater Development and Agrarian Change in Gujarat*. New Delhi: Oxford University Press.

Fatah, S. 2013. 'In India, Water and Inequality Are Intertwined'. The GroundTruth Project, 4 March. https://thegroundtruthproject.org/in-india-water-and-inequality-are-intertwined/ (accessed 21 December 2021).

FnBnews.com. 2019. 'India Ranks 120 among 122 Countries in Water Quality Index; 70% Contaminated'. 11 December. http://www.fnbnews.com/Top-News/india-ranks-120-among-122-countries-in-water-quality-index-70-contaminated-53487 (accessed 21 October 2021).

Hindustan Times. 2019. 'Nine-year-old Dalit Girl Raped in U'khand: Cops'. 31 May. https://www.hindustantimes.com/india-news/nine-year-old-dalit-girl-raped-in-u-khand-cops/story-haAfcZddVyTxYYaaN8FciI.html (accessed 21 December 2021).

Hulko, W. 2009. 'The Time- and Context-Contingent Nature of Intersectionality and Interlocking Oppressions'. Affilia 24 (1): 44–55.

Irudayam, A., J. P. Mangubhai, and J. G. Lee. 2006. *Dalit Women Speak Out Caste, Class and Gender Violence in India*. New Delhi: NCDHR.

Jinnah, S. 2011. 'Climate Change Bandwagoning: The Impacts of Strategic Linkages on Regime Design, Maintenance, and Death'. Global Environmental Politics 11 (3): 1–9. doi: https://doi.org/10.1162/GLEP_a_00065.

Joshi, D. 2011. 'Caste, Gender and the Rhetoric of Reform in India's Drinking Water Sector'. *Economic and Political Weekly* 46 (18): 56–63.

———. 2015. 'Like Water for Justice'. *Geoforum* 61: 111–121.

Kaijser, A., and Kronsell, A. 2014. 'Climate Change through the Lens of Intersectionality'. *Environmental Politics* 23 (3): 417–433.

Kashwan, P. 2006. 'Traditional Water Harvesting Structure: Community behind "Community"'. *Economic and Political Weekly* 41 (7): 596–598.

Kashwan, P. and V. Lobo. 2014. 'Of Rights and Regeneration: The Politics of Governing Forest and Non-Forest Commons'. In *Democratizing Forest Governance*, edited by S. Lele and A. Menon. New Delhi: Oxford University Press, 349–375.

Kashwan, P., L.M. Maclean, and G.A. García-López. 2019. 'Rethinking Power and Institutions in the Shadows of Neoliberalism: (An Introduction to a Special Issue of World Development)'. *World Development* 120: 133–146. https://doi.org/10.1016/j.worlddev.2018.05.026.

Khandekar, N., G. Gorti, S. Bhadwal, and V. Rijhwani. 2019 'Perceptions of Climate Shocks and Gender Vulnerabilities in the Upper Ganga Basin'. *Environmental Development* 31: 97–109. https://doi.org/10.1016/j.envdev.2019.02.001.

Krishnaraj, M. 2011. 'Women and Water: Issues of Gender, Caste, Class and Institutions'. *Economic and Political Weekly* 46 (18): 37–39.

Kulkarni, S. 2011. 'Women and Decentralised Water Governance: Issues, Challenges and the Way Forward'. *Economic and Political Weekly* 46 (18): 64–72. http://www.jstor.org/stable/41152345.

Kulkarni, S., S. Ahmed, C. Datar, S. Bhat, Y. Mathur, and D. Makhwana, 2008. *Water Rights as Women's Rights? Assessing the Scope for Women's Empowerment through Decentralised Water Governance in Maharashtra and Gujarat.* SOPPECOM, Utthan, and TISS. http://hdl.handle.net/10625/42606 (accessed 28 December 2021)._

Mansuri, G., and V. Rao. 2012. *Localizing Development: Does Participation Work?* Washington, DC: World Bank Publications.

Mehta, L. 1997. 'Water, Difference and Power: Kutch and the Sardar Sarovar (Narmada) Project'. Institute of Development Studies at the University of Sussex, Brighton, England.

———. 2016. 'Why Invisible Power and Structural Violence Persist in the Water Domain'. *IDS Bulletin* 47 (5): 31–42.

Moench, M. 1998. 'Allocating the Common Heritage: Debates over Water Rights and Governance Structures in India'. *Economic and Political Weekly* 33 (26): A46–A53. http://www.jstor.org/stable/4406930.

Moon, V., P. T. Borale, B. D. Phadke, S. S. Rege, and Daya Pawar, eds. 2014. *Dr. Babasaheb Ambedkar's Writings and Speeches*, Vol. 5. New Delhi: Dr. Ambedkar Foundation. Ministry of Social Justice and Empowerment, Government of India.

Narake, H., M. L. Kasare, N. G. Kamble, and A. Godghate, eds. 2003 *Dr. Babasaheb Ambedkar's Writings and Speeches*, Vol. 17, Part 1. Education Department, Government of Maharashtra.

Naz, F. 2015. 'Water, Water Lords, and Caste: A Village Study from Gujarat, India'. *Capitalism Nature Socialism* 26 (3): 89–101.

Nightingale, A. J. 2011. 'Bounding Difference: Intersectionality and the Material Production of Gender, Caste, Class and Environment in Nepal'. *Geoforum* 42 (2): 153–162.

Pal, S. 2019. 'Overlooked Correlation between Climate Change and Social Exclusion'. *Newsclick*, 11 July. https://www.newsclick.in/overlooked-correlation-climate-change-social-exclusion (accessed 21 December 2021).

Prakash, A. and R. K. Sama. 2006. 'Social Undercurrents in a Water-Scarce Village'. *Economic and Political Weekly* 41 (7): 577–579.

Punetha, P. 2018. 'Young Girls Forced to Skip School in Uttarakhand during Periods as Temples Fall on Path'. *Times of India*, 28 November. https://m.timesofindia.com/city/

bareilly/young-girls-forced-to-skip-school-during-periods-as-temples-fall-on-path/amp_
articleshow/66836940.cms (accessed 28 December 2021).

Ranjan, P., S. Kazama, and M. Sawamoto. 2006. 'Effects of Climate Change on Coastal Fresh Groundwater Resources'. *Global Environmental Change* 16 (4): 388–399.

Rao, N. D., J. Min, and A. Mastrucci. 2019. 'Energy Requirements for Decent Living in India, Brazil and South Africa'. *Nature Energy* 4 (12): 1025–1032. https://doi.org/10.1038/s41560-019-0497-9.

Reuters. 2019. 'In Drought-hit Delhi, the Haves Get Limitless Water, the Poor Fight for Every Drop'. *India Today*, 7 July. https://www.indiatoday.in/india/story/in-drought-hit-delhi-the-haves-get-limitless-water-the-poor-fight-for-every-drop-1563827-2019-07-07 (accessed 21 December 2021).

Roth, D., M. Zwarteveen, K. J. Joy, and S. Kulkarni. 2018. "Water Governance as a Question of Justice: Politics, Rights, and Representation." In *Water Justice*, edited by R. Boelens, T. Perreault, and J. Vos. Cambridge: Cambridge University Press. 43–58.

Scroll.in. 2018. 'Haridwar: Dalit Man Killed Allegedly over the Use of a Water Canal, Two Arrested'. 13 May. https://scroll.in/latest/878835/haridwar-dalit-man-killed-allegedly-over-the-use-of-a-water-canal-two-arrested (accessed 21 December 2021).

Sengupta, S. 2019. 'Chennai, an Indian City of Nearly 5 Million, Is Running Out of Water'. *New York Times*, 21 June. https://www.nytimes.com/interactive/2019/06/21/climate/chennai-india-water-shortage-images.html (accessed 21 December 2021).

Seymour, M. 2020. 'Understanding India's Dropout Rates for Girls'. *Borgen Magazine*, 25 September. https://www.borgenmagazine.com/indias-dropout-rates/ (accessed 21 December 2021).

Singh, V. 2019. 'Water and Hopes Down to a Trickle in Uttarakhand's Hilly Areas as Residents Ignored by State Abandon Their Homes'. *Firstpost*, 8 February. https://www.firstpost.com/india/water-and-hopes-down-to-a-trickle-in-uttarakhands-hilly-areas-as-residents-ignored-by-state-abandon-their-homes-6052751.html (accessed 21 December 2021).

Sirimane, M. and N. Thapliyal. 2020. 'Migrant Labourers, Covid-19 and Working-class Struggle in the Time of Pandemic: A Report from Karnataka, India'. *Interface: A Journal on Social Movements* 12 (1): 164–181.

Subramanian, M. 2019. 'India's Terrifying Water Crisis'. *New York Times*, 15 July. https://www.nytimes.com/2019/07/15/opinion/india-water-crisis.html (accessed 21 December 2021).

Tata-Cornell Institute for Agriculture and Nutrition. 2019–2020. *Tata-Cornell Institute Annual Report, 2019–20*. Ithaca, NY: Tata-Cornell Institute for Agriculture and Nutrition. https://tci.cornell.edu/wp-content/uploads/2021/04/TCI_Annual_Report_2019-20_Web.pdf (accessed 21 December 2021).

Thompson, J. A. 2016. 'Intersectionality and Water: How Social Relations Intersect with Ecological Difference'. *Gender, Place and Culture* 23 (9): 1286–1301.

Thorat, S. 2020. 'With the Sorry State of Dalits Still Evident in Data, the Judiciary Needs to Continue to Uphold Their Rights'. *The Hindu*, 24 February. https://www.thehindu.com/opinion/op-ed/batting-for-the-downtrodden/article30897192.ece (accessed 21 December 2021).

Upadhyay, V. 2011. 'Water Rights and the "New" Water Laws in India: Emerging Issues and Concerns in a Rights Based Perspective'. In *Water: Policy and Performance for Sustainable Development: India Infrastructure Report*, edited by P. Tiwari and A. Pandey. New Delhi: Oxford University Press, 56–66.

Varua, M. E., J. Ward, B. Maheshwari, S. Dave, and R. Kookana. 2018. 'Groundwater Management and Gender Inequalities: The Case of Two Watersheds in Rural India'. *Groundwater for Sustainable Development* 6: 93–100.

Vajpeyi, A. 2016. 'Ambedkar and the Struggle for Womens Equality'. *ANTYAJAA: Indian Journal of Women and Social Change* 1 (1): 5–9. doi:10.1177/2455632716645966.

Waghmare, A. 2016. 'How Water Inequality Governs Drought-hit Maharashtra'. *IndiaSpend*, 31 May. https://archive.indiaspend.com/cover-story/how-water-inequality-governs-drought-hit-maharashtra-24377 (accessed 21 December 2021).

Watkins, K. 2006. 'Human Development Report 2006: Beyond Scarcity: Power, Poverty and the Global Water Crisis'. New York: United Nations Development Programme.

Zwarteveen, M. Z., and R. Boelens. 2014. 'Defining, Researching and Struggling for Water Justice: Some Conceptual Building Blocks for Research and Action'. *Water International* 39 (2): 143–158.

Connected Worlds by Anupriya

Realizing Climate Justice through Agroecology and Women's Collective Land Rights

Ashlesha Khadse and Kavita Srinivasan

Introduction

The feminization of agriculture, or the sharp increase in the number of women in farming, is the result of a deep and ongoing agrarian crisis. Some scholars have more aptly named this phenomenon the 'feminization of the agrarian crisis' to capture how the ongoing agrarian crisis places a greater burden on women farmers than it does on their male counterparts. Patriarchal norms and attitudes prevent women from owning and controlling land, and women from marginalized castes and classes are the most disadvantaged (Pattnaik et al. 2018). Over 70 per cent of women in rural India are engaged in farming, but since the majority do not formally own land, they are not officially recognized as farmers and are instead considered as 'farm helpers' (Agarwal 2021). Given the substantial inequalities that affect women's ownership of and control over land, they cannot avail the benefits of land ownership – economic security, social status, and state support, among others.

This chapter looks at climate justice in the context of women in agriculture. Climate change and gender inequalities are deeply intertwined. Governments and civil society actors have launched various programmes aimed at climate resilience and adaptation in agriculture. However, when analysed through the lens of climate justice, these efforts do not always promote social equity. On the contrary, in some cases, mainstream climate solutions threaten women's land rights and farm-based livelihoods.

Using the novel framework of agrarian climate justice, which combines ideas from agrarian justice and climate justice, we explore women's land rights within agroecology programmes in India. We argue that advancing women's collective land rights through climate initiatives can achieve the twin aims of climate resilience and agrarian justice. We focus on agrarian land and do not look at forest lands, which, although equally important, are outside the scope of this chapter. Drawing from feminist scholars' work on intersectionality, we emphasize the importance of an intersectional understanding of the differences between women based on intersecting identities of caste, class, age, education, and marital status, among others (Lutz, Herrera Vivar, and Supik 2011). Such an understanding is important to ensure that climate policies reduce, instead of reproduce, inequalities.

Globally, peasant women's rights have received increasing attention from social movements concerned with agrarian and climate justice, which often overlap. One example is the global peasant movement La Via Campesina (LVC), for which the commitment to peasant women's rights was born as a result of women demanding and gaining leadership roles and space within the movement. LVC's emerging concept of popular peasant feminism recognizes structural causes of gender inequality and peasant women's rights to decision-making and resources, particularly land (Val et al. 2019). In India, the Mahila Kisan Adhikaar Manch (MAKAAM), the women farmers' rights platform, came into being because women had become invisible in agrarian policy and were not recognized as farmers despite their predominant role in agriculture. MAKAAM has been working to secure women farmer's rights and entitlements to receive equal support from the state, particularly in matters of land access and ownership. Both these networks see ecological approaches like agroecology as key elements in their feminist vision of a sustainable and just world. This chapter substantiates the conversation on agrarian climate justice through the perspectives of two Indian women farmers' collectives working in Tamil Nadu and Kerala.

In the next section, we introduce the framework of agrarian climate justice. This is followed by background information about women's land rights in India. We then present our two case studies, which are the Tamil Nadu Women's Collective (TNWC) and Kudumbashree in Kerala; both organizations seek to integrate women's livelihoods and collective land access with agroecology. We then offer key insights based on an analysis of the case material and reflect on the prospects of agrarian climate justice before concluding.

Conceptual foundations and methods

Climate change-related politics are increasingly linked to land, with an ongoing contest between grassroots actors like small farmers, indigenous peoples, and women on one hand and powerful market actors on the other. The link between land rights and climate interventions has been explored through the novel concept of 'agrarian climate justice' proposed by Borras and Franco (2018), which combines the principles of agrarian justice with climate justice. The concept of agrarian justice is related to struggles for recognition of land rights and redistribution and restitution of land, particularly for dispossessed groups like women. Climate justice is about equality and justice in the distribution of responsibility, impacts, and benefits accruing from solutions to climate change. Land is central to both climate justice and agrarian justice, and agrarian climate justice advocates the linking of social movements and the analyses of policies and programmes related to them.

Agrarian climate justice differentiates climate interventions promoted by the two sets of actors – market actors and grassroots actors – from their effect on land politics. Socially just land policies work towards regenerating nature while recognizing, redistributing, and returning land to the dispossessed (Borras and Franco 2018). In contrast, climate projects led by market-based actors and market processes threaten to dispossess rural populations of their lands to facilitate continuous capital accumulation. This approach tends to strengthen landed classes and agribusinesses while obfuscating redistribution. An example from India is the large-scale solar farms that have displaced vulnerable communities and facilitated the appropriation of village commons by renewables promoters (Yenneti, Day, and Golubchikov 2016). Land acquisition and popular movements against it have been well documented, but because of the lack of formal land titles for women, much of this discourse leaves out issues related to women farmers, which are linked to discussions on gender justice within climate justice.

Currently, gender justice in climate justice literature falls into three broad categories (Michael et al. 2019). The most prominent one highlights the gendered impacts of and vulnerabilities to climate change. The second category underlines the vital role women play in conserving the environment and promoting sustainability owing to their differential knowledge, roles, and stakes in ecological preservation. For example, Agarwal (2010) shows that in community-managed forests of India and Nepal, women are more responsible for firewood and fodder while men are more interested in timber. Such differing interests gave women a greater stake in forest preservation. The third category, building on the first two, advocates increasing the participation of women in decision-making processes and governance as a means to reduce gender injustice. Scholars point out that while these categories are all important, such conceptualizations also lead to problematic narratives around gender, deflecting attention from inequalities and power relations. For one, there is a tendency to portray women as a homogenous group and as vulnerable victims

(Crease, Parsons, and Fisher 2018). Second, they encourage an instrumental view of women – seeing them as responsible for making their families climate-resilient – leading to policies and practices that facilitate a feminization of both vulnerability and responsibility (Bendlin 2014).

Recent feminist scholarship has argued for a deeper intersectional analysis of climate interventions (Crease, Parsons, and Fisher 2018). An intersectional analysis considers inequalities not only between but also within genders, as depending on how they are situated, women have different vulnerabilities and adaptive capacities. Furthermore, it considers multiple identities working together to construct power or powerlessness. Gender intersects with other identities like caste, class, race, physical ability, and sexual orientation, among others. Not conducting such an intersectional analysis can lead to climate policies and practices that worsen rather than reduce inequalities, prompting gendered climate injustices.

Using Borras and Franco's (2018) framework for agrarian climate justice, gender-just climate interventions must accommodate recognition, redistribution, and restitution of land and other resources that are crucial for sustaining women-led agroecology. An intersectional approach steers the discourse away from the homogenization and essentialization of women and instead highlights the differences between women based on identities like caste, class, education, and marital status. Such a gender-focused intersectional analysis is a unique contribution to Borras and Franco's emerging framework for agrarian climate justice, which does not examine the question of women farmers' land rights.

Recognition entails acknowledging women's right to land; in India, this includes Dalit, Adivasi, or poor peasant women. Indeed, older women, single women, or women with disabilities within each of these social groups occupy an even more disadvantageous position. Redistribution of land to women is urgent where the means of production, especially land in rural areas, are monopolized by a few; in India, redistribution is a particularly pressing need among women who experience injustices borne of the intersection of multiple disadvantages. Restitution is relevant to those who have lost their land because of corporate or other types of resource grabs. It is also applicable to women – for example, widowed or divorced women – who may have lost land titles to other family members despite legal provisions to the contrary; in India, this includes Adivasi women whose forest commons have been grabbed for dams, mining, and other forms of resource extraction. Each of these dimensions must be examined from an intersectional perspective, with a focus on the recognition, redistribution, and restitution of the right to land for those individual or groups of women who hold a marginalized position, such as women from historically landless castes and classes.

To investigate the questions surrounding intersectional agrarian climate justice, we review policy documents, programmatic reports, non-governmental organization (NGO) reports, and social movements focused on two case study sites – the TNWC in Tamil Nadu and the state-wide Kudumbashree programme in Kerala. Additionally, we include information from qualitative interviews with leaders of the TNWC and scholars and activists of MAKAAM. We also draw from our association with the Karnataka chapter of MAKAAM, which has deepened its work related to women's collective farming efforts and organized various conversations and meetings on the topic with government officials, activists, and scholars, including the authors.

Background: land inequality and feminization of the agrarian crisis in India

Feminist scholars argue that landlessness is one of the most significant causes of female oppression in India (Agarwal 1995, 2003). Land access can provide both direct advantages like the ability to farm and indirect advantages that can take several forms, such as increasing bargaining power within and outside the household, enhancing social status, allowing access to state support, and encouraging the recognition of women as farmers. Additionally, there is some evidence that children of women with land tend to have better educational and health outcomes (Landesa 2012). Individual ownership also enables women to participate in credit markets using their land as collateral, but this comes with the dangers of land alienation and entrenches patterns of financialization (Collins 2019).

One important challenge in understanding women's land control is the lack of gender-disaggregated data in relation to land ownership since land title records are not digitized and most states do not collect data by gender (Swaminathan 2013). The only gender-segregated data that come close and allow some approximation are on operational land holdings (Table 10.1). An operational land holding is used wholly or partly for agricultural production and functions as one technical unit regardless of the title, legal status, farm size, or location. Agricultural census data from the Government of India reveal that 73.2 per cent of rural women workers are engaged in agriculture but that women control only 13.96 per cent of operational land holdings (Ministry of Statistics and Programme Implementation 2017). Moreover, even if women control 13.96 per cent of the land, it does not mean that they own this land since the data on operational land holdings do not account for title or ownership. There are significant gaps between women's legal rights and their actual inheritance of land and between the limited ownership rights women enjoy and their effective control over land. The gaps are mainly due to (*a*) gendered identities and social

Table 10.1 Operational land holdings of women from different social classes in India

	Number of operational holdings	Area operated (hectares)	Percentage of total (no. of holdings)
Men (all social groups)	125,751	137,784	85.80
Women (all social groups)	20,439	18,493	13.96
Women (Scheduled Castes)	2,329	1,584	1.50
Women (Scheduled Tribes)	1,612	1,984	1.10
Total (all social groups; men and women)	146,454	157,817	100.00

Source: Agricultural Census (2015–2016).

norms, which often restrict women's ability to articulate and exercise their right to inherit land, and (*b*) institutional practices, which are based on conventional male-dominated understandings of land ownership and inheritance (Sircar and Pal 2014).

The data point to skewed land ownership patterns both between genders and among women from different social classes. The data in Table 10.1 show that women from Scheduled Castes (SCs), the official term for Dalits, control only 1.5 per cent of the total land holdings in India, while women from Scheduled Tribes (STs), or Indigenous Peoples, control 1.1 per cent. Dalit women not only control less land than women from the so-called higher Hindu castes but are also mostly involved in agriculture as labourers rather than cultivators; this has a negative impact on their status. Other studies show that single women (unmarried, divorced, abandoned, or widowed) are the most vulnerable even within these social groups (Sircar and Pal 2014). These data and research findings make an important case for using an intersectional approach to study land access, as they highlight the social positions of women depending on their identities.

Recent structural changes in Indian agriculture have led to the increasing feminization of agriculture – women are participating in agriculture in larger numbers, as is evident in the rise in the percentage of land holdings operated by women between the last two agricultural censuses (Table 10.2). However, somewhat counterintuitively, the increasing feminization of agriculture is not necessarily linked to women farmers' empowerment (Pattnaik et al. 2018). Feminization is driven by an ongoing agrarian crisis that has rendered farming unviable for men. Moreover, the outmigration of men towards more viable livelihood opportunities has resulted in the growing labour contribution of women in agriculture. This adds to the already heavy work burdens of women and to the further deterioration of women's working conditions. Although an increasingly large number of women manage household agriculture operations, instead of being recognized as owner cultivators, they are regarded as agricultural labourers. Therefore, despite women's role in farming, they remain invisible. The lack of land titles, for instance, prevents women from being recognized as farmers in governmental programmes, such as those meant to subsidize the distribution of farm inputs or to facilitate easy access to rural agriculture credit. The situation is slightly improved in south Indian states like Andhra Pradesh, Tamil Nadu, and Kerala, which tend to have better land rights for women as compared to the rest of the country (Table 10.2). Nevertheless, the extent of overall landlessness is higher in all three states than in the rest of the country, with landless households at 73.41 per cent in Tamil Nadu, 72.50 per cent in Kerala, and 73.37 per cent in Andhra Pradesh, according to the latest available data (Ministry of Rural Development 2011). The struggle to secure women's land rights thus needs to be viewed in the context of the larger struggle for land rights for the landless poor.

The data in the table show standard land holding patterns in India, with a focus on operational land holdings. This unit of analysis has severe limitations, as it disguises problems like the fragmentation of operational holdings. It also does not take into account the ways in which landless women access land: increasingly, women are doing so collectively in some states. Given that the majority of India's rural women are landless, they encounter severe barriers, such as a lack of resources like land, inputs, capital, and skills, among others. Although institutions like the World Bank promote individual land rights within a liberal market-based framework, feminists note that without addressing broader social, political, and economic structures, individual land titles tend not to work for women (Jackson 2003). For instance, providing individual land titles without complementary support like inputs, training, credit, and culturally aware implementation will not result in any productivity gains, nor will it have transformative potential for gender relations. Contrarily, group farming

Table 10.2 Operational holdings of women and women cultivators by state in 2010–2011 and 2015–2016

	Percentage operated by women	
	2010–2011	2015–2016
Andhra Pradesh	22.10	30.09
Kerala	15.00	29.38
Tamil Nadu	16.60	21.02
All India	10.90	13.90

Source: Agricultural Census (2015–2016).

provides viable support for women to overcome such constraints by increasing their bargaining power and empowering them to pool resources, especially finance and land.

Women's land inheritance is governed by national laws, like the Hindu Succession (Amendment) Act, 2005 (HSAA), which ensures an equal share in ancestral property for men and women. HSAA was a significant move towards gender equality since land tenure rights were heavily biased against women in India before 2005. However, 15 years after its enactment, the ground reality is that women still do not inherit land on an equal basis with men. There are both formal and informal barriers to the implementation of HSAA and to the protection of women's right to land inheritance. Informal barriers include patriarchal pushback within the family such as resistance from brothers and parents, and cultural practices like dowry, because of which parents prefer to give dowry to their daughters and gift land to their sons (Landesa 2013). Formal barriers include lack of awareness and commitment among village councils and local land revenue staff who are meant to help enforce the act. Moreover, complicated procedures and administrative systems undermine women's ability to benefit from the law. Additionally, the HSAA does not apply to about 24 per cent of India's population comprising Muslims and Christians, who follow their own customary laws and have also traditionally excluded women from land ownership (Sircar 2016).

Tenancy laws, including land reform laws that impact women's access to land, are governed by states in India, so they vary across the country. All states have enacted reforms regarding the rights of tenants, labourers, and other farmers, but most have not accommodated women's land rights in a meaningful way (Chowdhry 2017). Despite the existence of the HSAA, there is legal ambiguity in its application to agricultural land, which falls under states' authority. Indeed, states have often overridden the HSAA with state-level land laws for agricultural land. One such example is the Uttar Pradesh Zamindari Abolition and Land Reforms Act, 1950, which discriminates married from unmarried daughters, as, in most cases, the former cannot inherit land (Mishra 2019). However, a 2019 Supreme Court decision has settled that agricultural land can be legislated by both central and state authorities, thus opening the doors for Hindu women to get succession rights to agricultural land under the HSAA (Supreme Court of India 2019). But in most cases, even if land records contain the name of a woman, the land is effectively controlled by male members of the family (Sircar and Pal 2014). Besides, inheritance is only possible for women from landed families. Women can also access land through government redistribution, land purchase, or by leasing directly from landowners. However, land redistributions have not historically worked in favour of women because of gender bias in state-led land reforms (Haque and Lekshmi Nair 2014). Moreover, purchasing is not easy, given land scarcity, prohibitive costs, and cultural norms that prevent women from accessing the necessary finances.

Under these conditions, land leasing has turned into an important path for women to access land, particularly via landless women's collectives formed by civil society groups or government institutions working with women from marginalized backgrounds (Agarwal 2003). Yet, the dynamics of group leases and agroecological enterprises remain inadequately documented and analysed. Next, we investigate two illustrative cases to inform our analyses in this chapter.

Case studies: group approaches to women's land access in two south Indian states

We present two programmes that promote climate-resilient agroecological farming while strengthening landless women's access to land, mainly through collective farming. We selected these cases because they offer unique perspectives on women's group farming and agroecology. Kudumbashree is an emblematic success story, achieved in part due to progressive policy interventions. TNWC is a case where policies on women's land access and agroecology are weak but women's movement efforts are prominent. Pragmatic reasons, such as having contact with these groups and being able to access programmatic documents easily, also played a part in our choices.

Tamil Nadu Women's Collective[1]

In Tamil Nadu, high levels of landlessness coupled with neoliberal reforms have led to a repurposing of agricultural land for non-agricultural uses. This has restricted the availability of arable land for landless women, particularly those from marginalized communities (Murthy 2017). In 2020, when we wrote this chapter, there was no large-scale organic farming programme in the state, although activists were demanding an agroecology policy. Tamil Nadu has a few programmes for rural women, most of which tend to focus on credit and livelihoods. Mahalir Thittam is a women's self-help group (SHG) building and poverty alleviation programme that operates in both urban and rural areas targeting women from poor households. The Tamil Nadu Rural Livelihoods Mission (TNRLM) is a livelihood-focused poverty alleviation programme linked to the Indian government's National Rural

---

[1] This information is based on several interviews the authors conducted with Sheelu Francis of the TNWC between June and August 2020.

Livelihoods Mission (NRLM), which includes some support for sustainable agriculture. The NRLM promotes agroecology to enhance women's livelihoods and climate resilience. In particular, one of the NRLM's more recent programmes, Mahila Kisan Sashaktikaran Pariyojana (MKSP, translated as 'Women Farmers' Empowerment Programme'), focuses on women and agroecology and is being implemented through the TNRLM. The MKSP programme aims to support 42,359 women in undertaking agroecological methods; however, there is no specific focus on landless women (Murthy 2017).

Women's land access in the state is due less to policy and more to do with the self-led initiatives of SHGs and the presence of a strong women's land rights movement that has helped landless women file petitions with the state to access unused government lands (Murthy 2017). A key member of the women's land rights movement in Tamil Nadu is the TNWC, a state-level federation of women's groups founded in 1994. With a membership of over 150,000 women, the TNWC is spread over 16 districts in Tamil Nadu. In its initial years, the TNWC provided counselling and legal aid to women who were victims of sexual violence, particularly caste-based sexual violence, which commonly arises in conflicts with landlords. Over time, the organization has expanded its activities to include sustainable solutions to food security and health. In this context, a focus on agroecology and land access has become one of the key pillars of the collective.

The TNWC organizes women into SHGs called *sangam*s. These *sangam*s engage in group savings to improve women's financial security and access to credit.

TNWC leaders note that women come together in groups to share farming resources, particularly land. Most of the TNWC's members are Dalits and tend to be either landless labourers or cultivators of small plots of land. Many of the women are single – either widowed, abandoned by partners, or unmarried – who single-handedly shoulder the responsibility of running their households. Most have no education. The women face discrimination for being Dalit and single, and rarely have access to land or other types of support from the government. Less than 10 per cent of TNWC members have land titles to their name. But the TNWC recognizes women's fundamental right to land and provides political education for women around this right. Some of the *sangam* members have been approached by state agencies to join state programmes. However, as an NGO, the TNWC does not have any formal role in the Mahalir Thittam or the TNRLM. Sheelu Francis of the TNWC points out that the TNRLM does not provide any land access support, which leaves out landless women, who tend to unify under social organizations like the TNWC. The TNWC assists 81 women's groups consisting of 715 members in total to engage in group farming over 91.74 acres.

The TNWC supports land access in a number of ways. It assists *sangams* in approaching the local government to make public land redistribution claims. But government officials are often apathetic, which causes delays and disappointments. The TNWC advocates for collective as opposed to individual land grants. The latter do not guarantee women control over the land and do not prevent land from being bequeathed to sons and thus taken out of women's hands. It encourages the *sangams* to use their own savings to purchase land from the market but notes that the high cost of land is prohibitive. Indeed, the most common way for women to access land is by leasing it. This is done via a lease agreement with a landowner who is oftentimes a woman, such as a widow who may have inherited, but does not cultivate, the land. Such single women are invited to become part of the group via a share-cropping arrangement. To minimize lease payments, the TNWC members split the costs and share a third of the produce with the landowner. This encourages landless women to make alliances with landed single or older women who cannot work on their land themselves. Such women are more easily able to enter into joint cultivation arrangements if they are the sole owners of their land rather than joint owners with their husbands.

Land leasing is often fraught with insecurity for women. Often, when landowners see the land improve after agroecological farming, they want it back for themselves. *Sangams* therefore prefer longer and formal leases, for at least five years, but most landlords prefer informal leases so that they can take the land back anytime; this practice is restricted under Tamil Nadu's land lease laws. The TNWC currently advocates for long-term secure land leases for women's *sangams* in cooperation with the Tamil Nadu government.

In addition to promoting access to land, the TNWC supports *sangams* with credit and training on saving and thrift activities. Members contribute at least ₹100 per month to their *sangam* – this is pooled to support joint farming activities and loans for members. The TNWC gives an initial loan or seed capital of ₹4,000 to each group to supplement the women's own investments. As institutional or even informal credit is usually unavailable to landless women, the seed capital helps to fill this gap. When returned, the funds are passed on to another group.

The TNWC trains *sangam* members in technical aspects of farming like crop selection, agroecology, water conservation, and seed saving. During such training sessions, participants discuss relevant topics like violence against women, women's land rights, sustainable diets, and climate change, among others. The TNWC has also designated one or two model farms in each of the 16 districts, which serve as demonstration and training facilities for newer groups. Some women's groups maintain seed banks that facilitate the sharing of seeds within the network. Given that drought is a serious problem in many of the villages, millet-based farming is encouraged; this has helped the women's groups adapt to dry conditions while contributing to household food security.

The TNWC's work has led to several positive outcomes for its members. The building of strong social networks for women farmers has helped women resolve a number of problems and fosters confidence in them. The groups facilitate peer learning and the pooling of risks related to crop failures due to drought. *Sangams* also help with food security and access to credit for landless women, many of whom face absolute poverty. Further, growing food through *sangams* and having an assured source of income greatly enhances food security for families.

The women farmers share that group farming has brought them more respect in their community. In the initial days of the group's formation, community members and upper-caste landlords subjected them to scrutiny, gossip, and ridicule. However, this has changed, as the women have persisted and succeeded in farming. Now, male farmers even ask them for seeds and farming advice. The women have also had a positive impact on youngsters who grow up seeing their mothers and sisters as role models.

Kudumbashree, Kerala

Kerala has long been known for its relatively successful abolition of feudalism and land reforms that were effected in 1970s. Yet the state's land reforms have reinforced patriarchal norms by identifying the marital family as the unit of reforms (Kodoth 2009). More than a fourth of those who lost land as a result of land reforms in the state were widows (Haque and Lekshmi Nair 2014). The post-reform period led to a decline in women owner cultivators and a consequent rise in women's farm work within the context of the family as well as the outmigration of men. Changes in dowry practices in recent times have also affected women's land inheritance. Daughters previously got land as part of their inheritance, mainly in matrilineal families, but as dowry practices have gained wider acceptance, parents prefer to give movable property or cash, which can be invested elsewhere. This has further reduced women's land access through inheritance (Kodoth 2004).

One key path for marginalized women in Kerala to land access is the state's poverty alleviation and livelihoods programme. Kudumbashree was initiated in 1998 under the NRLM's state-level programme, the State Poverty Eradication Mission. The aim of the Kudumbashree Mission is to eradicate poverty through various economic enterprises, of which group farming is an important component (Agarwal 2019). Kerala has been undergoing a rapid decline in agriculture resulting from the outmigration of men and waning interest in agriculture among traditionally cultivator families, which has led to large areas of land becoming fallow (Kudumbashree 2020). Land leasing is also banned in the state, which has exacerbated the problem of fallow lands. Kudumbashree has used this as an opportunity to get more land into the hands of women's joint farming groups.

Women's group farming and economic enterprises in Kerala are supported by the institutional structure of the Kudumbashree programme, which is founded on three main pillars (Agarwal 2019). The first pillar is the Kudumbashree Mission, or the state poverty alleviation programme, run by government officials from several departments. The second is the grassroots members' network of Kudumbashree called the Kudumbashree community network – this is made up of all the women members who participate in the programme. At the lowest neighbourhood level, the basic unit of organization is called the neighbourhood group (NHG), which is similar to SHGs elsewhere. NHGs are small groups of 5–10 women who live close to each other and come together to initiate group economic enterprises, including group saving and thrift. Membership in NHGs is limited to one woman per family; Kudumbashree ensures that all poor families join NHGs. One of the key interventions of NHGs is collective farming by women farmers. The NHG is the basic unit of intervention for Kudumbashree and other government programmes. For example, the state government's agroecological farming programmes are disseminated and implemented via NHG networks. The network of NHGs is federated at the panchayat level, which means that all NHGs in a particular panchayat are registered as one autonomous organization with elected leadership. The third pillar of the Kudumbashree programme is the panchayat-level institution or the local government. The Kudumbashree community network mediates between the panchayat and the Kudumbashree Mission.

Kerala's commitment to agroecology started in 2014 when it created a state organic farming policy. More recently, in 2019, Kudumbashree initiated a climate resilience programme to turn 10,000 hectares into organic farming land and ensure certification in certain identified areas of all districts (Kudumbashree 2020). This programme is being implemented and scaled up via women's farming collectives. The farming groups are called joint liability groups (JLG); this name refers to the joint obligation of a group to pay debts. India's National Bank for Agriculture and Rural Development (NABARD) created JLGs to enable banking institutions to give out joint loans (Agarwal 2018). The JLGs, which comprise 4–10 women each and are embedded in the NHG network, are also federated at the cluster level; training and organic certification are implemented at this level. The number of JLGs leasing land has steadily increased from 26,499 in 2006–2007 to 65,601 in 2016–2017, and the area cultivated has increased from 17,370 hectares in 2006–2007 to 51,113 hectares in 2016–2017 (Abraham 2019).

Although Kerala banned the leasing of land to individuals through the Kerala Land Reform Act, 1963, following an amendment in 1969, it now allows land leasing to JLGs under the Kudumbashree programme using informal leases (Haque and Lekshmi Nair 2014). Local panchayat institutions support JLGs in identifying fallow lands and facilitating leases. However, since land leasing is officially banned, JLGs enter into informal leases. Usually, the leases are oral or written on paper, but informal and unregistered (Haque and Lekshmi Nair 2014). Banks recognize such informal leases in giving credit to JLGs. Clearly, a lack of formal leases does not prevent women from accessing state support for farming. However, the informal nature of the leases translates into tenurial insecurity. Kudumbashree women farmers often find that landowners are unwilling to negotiate longer than one-year term agreements. In one reported case, a landowner claimed state incentives in his own name and prevented the JLG from doing so (Abraham 2019). Fragmentation of land holdings is another problem and, as a result, many JLGs fail to find contiguous plots – the disparate lands are too small to cultivate individually.

Kudumbashree has an incentive structure to encourage JLGs to take up agroecological farming. However, JLGs are free to practise chemical farming or agroecological farming; they receive some incentive for both types, but they can secure additional incentives for the latter. The incentives are mainly aimed at achieving economies of scale necessary for the commercial viability of the group enterprises. There are two kinds of incentives. Area incentives are meant for women to lease fallow land. These incentives apply to a minimum area of 0.2 hectares. If a woman's group cultivates over 0.2 hectares of fallow land, they are eligible for a subsidy that equals 10 per cent of the total production costs incurred by the group. The second type of incentive is available to women who practise agroecological farming, which is certified by the local agriculture office. The women receive an additional 50 per cent of the 10 per cent area incentive. Thus, SHGs/JLGs that take up agroecological farming receive 1.5 times the area incentive.

A key factor in Kudumbashree's success is its wide institutional network (Agarwal 2018; Pammi and Malamasuri 2014). *Gram* panchayats provide inputs (seeds, fertilizers, and manure), basic infrastructure, machinery, irrigation facilities, and one-time land development grants for farming. The Kudumbashree community network provides support and training via agricultural universities and expert farmers within the network, whom they call 'master farmers'. The JLGs can tap into the Mahatma Gandhi National Rural Employment Guarantee Act (MGNREGA) labour pool to access workers for farming activities – local village councils coordinate this. There are multiple sources of loans; for instance, the state's primary agricultural co-operative societies (PACS) provide interest-free loans for selected crops. Nationalized banks and private banks provide crop loans at an interest rate of

7 per cent, of which Kudumbashree subsidizes 5 per cent. The programme facilitates market linkages with local and district markets. The JLGs sell their produce to each other and at a weekly market that Kudumbashree organizes to eliminate middlemen.

The key outcomes of the programme include improvements in the livelihoods of economically poor and socially marginalized women. Initially, most Kudumbashree collective farmers were landless labourers. They worked for wages and did not have land or access to credit. Today, the number of women taking up farming in the state has increased significantly (Abraham 2019). Women's food access has improved, as has their income from sales. Studies show that women's collective farms are considerably more productive than individual family farms (Agarwal 2018). The availability of fresh produce has also gone up significantly in the local markets where Kudumbashree women sell produce (Abraham 2019).

Analysis

Here, we analyse the extent to which the two state-level programmes serve the interests of women farmers and strengthen the grassroots practices of agroecology. We consider the advancing of these two aspects as contributing to the goals of agrarian climate justice. Viewing our case studies within an agrarian climate justice and intersectional framework then brings us to the question of whether the land-related policies and agroecology programmes in the two states have actually improved women's land rights and who are the women who have benefited. In our analysis, we focus on (*a*) state policies and programmes for agroecology, (*b*) state policies and programmes for improving access to land, (*c*) other vital resources (access to credit, training, and markets), and (*d*) informal sociocultural norms that interfere with the design and implementation of the programmes.

State policies and programmes for agroecology

Tamil Nadu and Kerala occupy contrasting positions when it comes to agroecology policies. Kerala has an ambitious plan to convert 10,000 hectares to organically cultivated land and a state organic policy that demonstrates its commitment to climate resilience. This strategy is being implemented via women's livelihoods programmes and women's collectives in the state. The focus on collective approaches and women's livelihoods is critical in enacting agrarian climate justice, as it ensures that marginalized groups benefit from agroecology programmes. Kudumbashree's universal coverage that is open to one woman from each family guarantees that vulnerable women benefit from these programmes.

Kerala's three-pillared institutional support system, which includes local government, the Kudumbashree Mission, and the community network, has been successful in facilitating sustainable collective farming and agroecological work. Support via national agroecology programmes like the MKSP is embedded into Kudumbashree's existing programme. An incentive structure to promote agroecology supports women's groups' practice of agroecological farming. The effort to bring fallow agricultural lands into the fold of agroecology is a key step towards both climate resilience and social justice. Meanwhile, Tamil Nadu does not have a specific agroecology programme despite a strong social movement promoting agroecology in the state. It is implementing the recent national-level MKSP programme for women's agroecology training as part of the TNRLM and Mahalir Thittam women's SHG network. However, the TNWC case shows that this programme does not necessarily reach landless women, who do not get support for land access. While the programme does try to include poor households, its lack of focus on land access for women limits its contribution to the advancement of agrarian climate justice, which requires a strong focus on land rights.

State policies and programmes for improving access to land

An analysis of land access policies in the two states reveals that of the various paths women can take to access land – inheritance, market purchase, government redistribution, and leasing – the first three have not benefitted women, particularly those from marginal castes. When it comes to the redistribution of public lands, women's collectives like the TNWC encourage local governments to allocate such land. However, these pursuits are often mired in bureaucratic processes and subject to the whims of individual government officials. There is no concrete law in either state to promote women's collective rights to public lands that have been set aside for redistribution. The large-scale conversion of agricultural lands for commercial and residential use in states like Tamil Nadu restricts the possibility of women's land rights while strengthening market actors' claims to common land.

The most common way in which resource-poor women in both our case studies access land is by leasing it. In both states, collective farming via land leasing has led to an increase in women's participation in agriculture as cultivators rather than labourers. In Kerala, although land leasing is banned, the state supports leasing by women's groups as a strategy to secure land access for women and reverse the expansion of fallow land. Local government institutions have helped women from the Kudumbashree network identify land and facilitate leasing along with access to credit. This shows the state's commitment to ensuring women's collective land rights.

The state's support seems to be critical to women successfully accessing land. In Tamil Nadu, land leasing is permitted, but there is no specific support for women farmers. Women face constraints in identifying and leasing land themselves. Organizations like the TNWC fill this crucial gap by bringing women landowners into women's collectives and negotiating longer leases and share-cropping arrangements, without renters having to pay rent in cash. Tamil Nadu should recognize and support such a pooling of resources by women who have land with others who can provide labour, as this can strengthen women's collective land rights and promote agrarian climate justice.

Women in both states face tenurial insecurity and there is a tendency for leasing arrangements to be informal, short, and insecure. This is often the result of landlords' reluctance to enter into formal lease agreements out of fear of losing control over the land or access to state subsidies, which are linked to land ownership. Women members report landlords wanting to cancel their leases after they see that agroecological farming improves the land. Similarly, Kudumbashree women note that it is difficult to get leases longer than one year. Formalizing tenancy laws so that women's collectives can get longer leases and the associated benefits of state programmes would help mitigate these difficulties. Our analyses of the two cases show that enacting laws is necessary but rarely sufficient to bring about transformative change.

While land purchase has not been a significant source of land access in either state, it is an important avenue for women's collectives that manage to accumulate funds. It is critical that state policy recognize not just individual women, but all forms of women's collectives, such as SHGs, JLGs, and cooperatives, as valid landowners. Further, as highlighted in our literature review, individual land titles promoted within a market-based framework do not necessarily mean that women control the lands, which have been subject to financialization via land and credit markets. Nevertheless, both individual and collective land titles are important to the state recognizing women's land rights.

Other vital resources

Land redistribution and land titling have been the focus of many land rights movements. However, feminist critics have pointed out that without a host of supportive mechanisms, giving out titles is not enough to make the land productive (Jackson 2003). In both cases, we find that land access is just the first step to making agroecology viable. NGOs, community organizations, and social movements provide vital resources for, and commitment to, such work. In Tamil Nadu, where there is little institutional support and the state is not committed to redistributive land justice for women, organizations like the TNWC help vulnerable and marginalized women to claim collective land rights. The TNWC also provides support in the form of technical training, inputs, seeds, and credit to help women practise agroecology. In Kerala, the three-tiered institutional support structure – Kudumbashree community network, Kudumbashree Mission, and local government institutions – provides an ecosystem of support to women's collectives, for instance, through land leasing, extension services, incentives, and marketing. Given that women need a variety of support to successfully practise collective agroecological farming, policy interventions should prioritize building these foundations. Further, the role of external actors like NGOs and grassroots movements must be recognized and rewarded in programme implementation.

Informal sociocultural norms that interfere with the design and implementation of programmes

Our case studies confirm that patriarchal sociocultural norms impede women in their efforts to practise collective agroecological farming. Male community members often ridicule women's efforts to farm independently, as it contradicts conventional and patriarchal understandings of gender roles and the caste position of women. Indeed, the TNWC women, many of whom are Dalits, faced resistance and ridicule from upper-caste landlords in their initial experiments with group farming. Caste conflicts around land are common, and the TNWC case shows that Dalit women are subjected to physical, sexual, and verbal violence when they try to assert their rights. Women farmers also encounter apathy from state officials when they approach them for land allocations; this also has to do with cultural attitudes around women's land ownership. In Kerala, the situation is different as the state's women's collectives are further along in their work and there is greater social acceptance of them. Their successes are also the result of strong state support for women's joint farming and better provisions for gender rights in the state. In areas of Tamil Nadu where women's group work has been ongoing for some time, the women find that men in the community are more supportive of their work.

These findings offer useful insights for programme design. Accounting for such sociocultural norms in programme design, building strong institutional support structures for women's collectives, sensitizing government officials, promoting women's movements, and hiring more women for relevant government positions can help address the barriers linked to conventional gender norms.

Conclusion

This chapter approached climate justice as it applies to women in agriculture. Using the agrarian climate justice framework, we argued that the twin aims of agrarian justice and climate justice must be addressed by advancing women's right to and control over land in climate resilience initiatives. We lookd at agroecology programmes in Tamil Nadu and Kerala, initiated by both state and movement actors, and analysed the prospects for advancing agrarian climate justice in policy and programmatic initiatives.

We highlighted the fundamental inequalities in women's ownership of and control over land, particularly concerning women from marginal castes and classes. Given the ongoing feminization of the agrarian crisis, which means that women are increasingly participating in farming as labourers rather than cultivators, the issue of their land rights is even more important. Our analyses showed that collective farming and collective land leasing offer significant benefits in terms of women's land rights and overall wellbeing. They further revealed four key areas within state policy that affect agrarian climate justice: (*a*) agroecology policies, (*b*) land access programmes for women, (*c*) other resources that help to make agroecology viable, like support from local NGOs/movements, access to credit, training, and marketing support, and (*d*) informal sociocultural norms that interfere with the implementation or design of gender-just agroecological programmes.

The two case studies provided a number of insights that could improve the prospects of agrarian climate justice in agroecology initiatives from the perspective of women farmers. One is a convergence of women's agroecology-based livelihood programmes with land access so that institutional support for both agroecology and land can be offered simultaneously. Many states have already made women's livelihood interventions via SHGs and are currently implementing various agroecology extensions through these interventions. Linking these with land redistribution or leasing, particularly through collective land access as is being done in Kerala, can greatly enhance the ability of poor and landless women to successfully practise agroecology and advance their land rights.

Leasing has become an important way for women's groups to access land, but these women face severe insecurity due to informal, insecure, and short leases. Formalizing tenancy laws to allow women's collectives secure land access could greatly improve women's collective land rights. Access to land is a key aspect, but not enough to ensure viable agroecological farming. A host of supportive measures are needed to make farming a success, including technical training, procurement, and links with local governments. Regressive sociocultural norms must also be tackled via sensitization and training, particularly within government offices at the local level. Social movement actors can provide critical support in programme design and implementation, given their vast experience, large community networks, and social justice vision.

Acknowledgements

We are grateful to the members, leaders, and staff of the Tamil Nadu Women's Collective and Kudumbashree for their critical work and willingness to share their experiences with us.

References

Abraham, Tresa D. 2019. 'Lease land Farming by Women Collectives: An Enquiry into Earnings of Kudumbashree Groups'. Occasional paper, Centre for Women's Development Studies, New Delhi.

Agarwal, B. 1995. *A Field of One's Own*. Cambridge: Cambridge University Press. doi:10.1017/cbo9780511522000.

———.2003. 'Gender and Land Rights Revisited: Exploring New Prospects via the State, Family and Market'. *Journal of Agrarian Change* 3 (2): 184–224.

———. 2010. *Gender and Green Governance: The Political Economy of Women's Presence Within and Beyond Community Forestry*. *Gender and Green Governance: The Political Economy of Women's Presence Within and Beyond Community Forestry*. New Delhi: Oxford University Press, doi:10.1093/acprof:oso/9780199569687.001.0001.

———.2018. 'Can Group Farms Outperform Individual Family Farms? Empirical Insights from India'. *World Development* 108: 57–73. https://www.sciencedirect.com/science/article/pii/S0305750X18300913.

———.2019. 'The Interplay of Ideas, Institutional Innovations and Organisational Structures: Insights from Group Farming in India'. ESID Working Paper 116, Manchester, https://papers.ssrn.com/sol3/papers.cfm?abstract_id=3430406#references-widget (accessed 16 July 2020).

———.2021. 'The Invisible Farmers'. *Outlook Magazine*, 1 February. https://magazine.outlookindia.com/story/india-news-the-invisible-farmers/304184 (accessed 30 June 2021).

Bendlin, L. 2014. 'Women's Human Rights in a Changing Climate: Highlighting the Distributive Effects of Climate Policies'. *Cambridge Review of International Affairs* 27 (4): 680–698.

Borras, S. M., and J. C. Franco. 2018. 'The Challenge of Locating Land-Based Climate Change Mitigation and Adaptation Politics within a Social Justice Perspective: Towards an Idea of Agrarian Climate Justice'. *Third World Quarterly* 39 (7): 1308–1325.

Chowdhry, P., ed. 2017. *Understanding Women's Land Rights: Gender Discrimination in Ownership*. Thousand Oaks, CA: Sage Publications.

Collins, A. M. 2019. 'Financialization, Resistance, and the Question of Women's Land Rights'. *International Feminist Journal of Politics* 21 (3): 454–476.

Crease, R. P., M. Parsons, and K. T. Fisher. 2018. 'No Climate Justice without Gender Justice: Explorations of the Intersections between Gender and Climate Injustices in Climate Adaptation Actions in the Philippines'. In *Routledge Handbook of Climate Justice*, edited by T. Jafry. London: Routledge, 359–377.

Haque, T. and J. L. Nair. 2014. 'Ensuring and Protecting the Land Leasing Right of Poor Women in India'. In *2014 World Bank Conference on Land and Poverty*. Washington DC: Landesa.

Jackson, C. 2003. 'Gender Analysis of Land: Beyond Land Rights for Women?' *Journal of Agrarian Change* 3 (4): 453–480.

Kodoth, P. 2004. 'Gender, Property Rights and Responsibility for Farming in Kerala'. *Economic and Political Weekly* 39 (19): 1911–1920.

———. (2009. 'Residual Farmers on Household Land? Women and Second Generation Concerns of Regulation in Kerala'. In *Gender Discrimination in Land Ownership: Land Reforms in India*, Vol. 11. New Delhi: Sage Publications Inc., 117–140.

Kudumbashree. (2020). 'Kudumbashree: Organic Farming'. http://www.kudumbashree.org/pages/670 (accessed 20 April 2020).

Landesa. 2012. 'Women's Secure Rights to Land: Benefits, Barriers, and Best Practices'. Landesa Issue Briefs. https://www.landesa.org/wp-content/uploads/Landesa-Women-and-Land-Issue-Brief.pdf (accessed 21 April 2020) .

———. 2013. *The Formal and Informal Barriers in the Implementation of the Hindu Succession (Amendment) Act 2005*. UN Women, www.unwomen.org.

Lutz, H., M. T.Herrera Vivar, and L. Supik. 2011. *Framing Intersectionality: Debates on a Multi-Faceted Concept in Gender Studies*. London: Ashgate.

Michael, K., M. K. Shrivastava, A. Hakhu, and K. Bajaj. 2019. 'A Two-Step Approach to Integrating Gender Justice into Mitigation Policy: Examples from India'. *Climate Policy* 20 (7): 800–814.

Ministry of Rural Development. 2011. *Socio-Economic Caste Census-2011*. https://secc.gov.in/statewiseLandOwnershipReport?reportType=Land%20Ownership (accessed 21 April 2020).

Ministry of Statistics and Programme Implementation. 2017. *Periodic Labour Force Survey (PLFS)*. Delhi. http://mospi.nic.in/publication/annual-report-plfs-2017-18 (accessed 21 April 2020).

Mishra, A. 2019. 'Hindu Women's Inheritance Right in Agricultural Property: Myth or Reality'. *SSRN Electronic Journal*. doi:10.2139/ssrn.3382304.

Murthy, R. K. 2017. 'Tamil Nadu Women's Land Rights in the Context of Neo-liberal Tamil Nadu'. In *Understanding Women's Land Rights: Gender Discrimination in Ownership*, vol. 13, edited by P. Chowdhry. Thousand Oaks: Sage Publications.

Pammi, R. K., and K. Malamasuri. 2014. "Joint Farming through Neighborhood Groups (NHG's) in Kerala: A Case of Kudumbaashree'. *Society for Scientific Development in Agriculture and Technology* 9: 1010–1013.

Pattnaik, I., K. Lahiri-Dutt, S. Lockie, and B. Pritchard. 2018. 'The Feminization of Agriculture or the Feminization of Agrarian Distress? Tracking the Trajectory of Women in Agriculture in India'. *Journal of the Asia Pacific Economy* 23 (1): 138–155.

Sircar, A. K. and S. Pal. 2014. 'What Is Preventing Women from Inheriting Land? A Study of the Implementation of the Hindu Succession (amendment) Act 2005 in Three States in India'. In *World Bank Conference on Land and Poverty*. Washington DC: The World Bank.

Sircar, A. 2016. 'Women's Right to Agricultural Land: Removing Legal Barriers for Achieving Gender Equality'. Oxfam India Policy Brief No 19. https://oxfamilibrary.openrepository. com/bitstream/handle/10546/618600/bn-womens-right-agricultural-land-india-070716-en.pdf?sequence=1&isAllowed=y (accessed 10 May 2020).

Supreme Court of India. 2019. *Babu Ram vs Santokh Singh*. Civil Appeal No. 2553. https://main. sci.gov.in/supremecourt/2018/40770/40770_2018_Judgement_07-Mar-2019.pdf (accessed 30 June 2021).

Swaminathan, M. 2013. 'Gender Statistics in India'. Indian Statistical Institute. http://www.fas. org.in/pages.asp?menuid=16 (accessed 21 April 2020).

Val, V., P. M. Rosset, C. Zamora Lomelí, O. F. Giraldo, and D. Rocheleau. 2019. 'Agroecology and La Via Campesina I. The Symbolic and Material Construction of Agroecology through the Dispositive of "Peasant-to-Peasant" Processes'. *Agroecology and Sustainable Food Systems* 43 (7–8): 872–894.

Yenneti, K., R. Day, and O. Golubchikov. 2016. 'Spatial Justice and the Land Politics of Renewables: Dispossessing Vulnerable Communities through Solar Energy Mega-projects'. *Geoforum*, 76: 90–99. https://www.sciencedirect.com/science/article/abs/pii/S0016718515303249.

CHAPTER 11

Conclusion

Pathways to Policies and Praxis of Climate Justice in India

Prakash Kashwan and Eric Chu

Introduction

Anil Agarwal and Sunita Narain published a remarkable report in 1991 in which they differentiated between 'survival emissions' and 'luxury emissions'. It would not be an exaggeration to say that no other report has had a comparable impact on global debates and scholarship on climate justice. This distinction between survival and luxury emissions has been central to some of the most important pieces of scholarship and advocacy on climate justice (Shue 1993). Based on this report, common but differentiated responsibilities (CBDR) became the defining feature of the Indian government's position in international climate negotiations (Jasanoff 1993). Despite having such a massive influence on international climate negotiations, the distinction between survival and luxury emissions is rarely referenced in domestic climate policy debates. Even as climate disasters, including cyclones, floods, and heatwaves, become more intense, there is limited public debate on climate action and policy in India (J. Das 2020). On the other hand, while there is robust scholarship on India's climate policy and action in the international arena, engagement with questions of domestic climate justice within Indian academia is quite sparse (Fisher 2015; Chu and Michael 2019). The potential for domestic injustices was apparent even in 1991 and was duly acknowledged in the same Centre for Science and Environment (CSE) report that made CBDR foundational to India's position in international negotiations:

> Can we really equate the carbon dioxide contributions of gas guzzling automobiles in Europe and North America *or, for that matter, anywhere in the Third World* with the methane emissions of draught cattle and rice fields of subsistence farmers in West Bengal or Thailand? Do these people not have a right to live? But no effort has been made in WRI's report to separate out the 'survival emissions' of the poor, from the 'luxury emissions' of the rich. (Agarwal and Narain 1991, 3, italics added for emphasis)

For a variety of reasons that require deeper inquiry, questions of domestic climate justice fell through the intertwined cracks of international climate change politics and sectoral silos that are endemic to both academic research and grassroots social movement organization (Gupta 2014). Many argued, quite appropriately, that the policy priority should be addressing issues of employment, food security, education, and primary healthcare for the poorest people in India and other countries in the Global South. However, it is not helpful to maintain this development-climate action dichotomy. It is quite well known that the climate crisis has only made poor people's lives even more precarious, further exacerbating deeply entrenched development inequities. Yet, as an Indian climate activist wrote sometime back, 'among many left friends, mention of global warming gets a blank look' (Adve 2007, 1002–1003). The parliamentary left in India continues to be too weak to make a difference, but the same cannot be said about other national political parties or India's celebrated civil society and social movements. It is evident that these social and political actors could do more to create broad-based coalitions to support more progressive domestic climate action and climate justice (Bidwai 2012). For the most part, social science scholarship could do more to challenge the undercurrents of 'climate nationalism' that run through debates on India's stance in international climate negotiations.

Each of the mechanisms outlined above – rooted in the specificities of politics, political economy, and scholarly analyses of India's climate position – have reinforced the continued neglect of climate justice within India's borders. As India focuses on smart technologies and modernist industrial growth, the agenda of responding to climate change related risks to local jobs, schools, health services, food, and shelter has fallen by the wayside. The primary motivation for this volume was to address this justice 'gap' between the worlds of climate policy research, scholarship, and activism. However, instead of presenting an all-encompassing abstract discourse, each of the chapters in the volume seeks to unpack climate justice debates in specific policy and programmatic areas – national and state climate action plans, emission inequalities, the transition away from fossil fuels, the anticipated transition to renewable energy, urban governance, access to drinking water, women's access to farmland and agroecology, caste injustices, and India's environmental and climate movements.

Each chapter demonstrates how broader processes as well as power, socioeconomic inequalities, and neoliberalism are entangled in the ongoing public debates, policy processes, and programme development relevant to climate action in India. Most writings on climate 'justice' mainly provide an understanding of the drivers, manifestations, and effects of injustices. However, the contributors in this edited volume go the extra mile to offer analyses that inform the pursuit of climate justice – they engage with policies, programmes, and mobilizations that contain in them the seedlings, or in some cases saplings, of climate justice. Since these analyses are based on in-depth engagements with sociopolitical contexts and institutional structures, they do not devolve into simplistic, one-size-fits-all, technocratic solutions. As the next section explains, each contribution in the preceding pages engages with a specific question, issue, or policy area, analysing the most important barriers to as well as the constituents of a just approach to climate action.

Key insights from the contributions in this volume

Much of climate policy literature on India presumes that reducing greenhouse gas emissions necessarily entails trading off the country's development interests. Haimanti Bhattacharya's innovative research on the potential links between state-level emissions and economic inequalities offers a major corrective against this assumption. She shows that this relationship was negative before the onset of economic reforms – that is, lower levels of economic inequality at the state level were associated with higher levels of carbon emissions before 1991. However, in the post-economic reform era, this relationship has turned positive – states with higher levels of economic inequality also have higher levels of carbon emissions. This suggests that in the post-economic reforms era, a few states have witnessed an increasing concentration of both wealth and emissions. This finding has two somewhat contradictory implications for climate justice. On the one hand, it means that India's emissions are now more highly concentrated among those who benefit from the status quo than they were before the onset of economic reforms. On the other hand, it also means that significant emission reduction is possible by regulating the activities of the richest 10 per cent of India's population. If the cost of these regulations is borne by this population, aggressive climate action will not produce regressive social outcomes, especially if sectors with multiplier effects, such as food production and freight transport, are protected against inflationary impacts.

Vasudha Chhotray's analysis of the state-led coal sector demonstrates that 'extractive regimes' – which are amalgams of political, institutional, and discursive apparatuses – circumscribe the possibilities of justice. Chhotray argues that similar regimes will shape renewable energy developments unless they include bottom-up political engagements with grassroots actors and networks. Chhotray's arguments find further support in Karnamadakala Rahul Sharma and Parth Bhatia's analysis of India's state-controlled power sector, which they characterize as 'gigantic' in scale. They argue that the continued concentration of power among political and economic elites in the transition to renewable energy systems can be disrupted if public policies link energy system choices to social justice goals and the redistribution of political power within Indian society. They caution against pinning one's hopes for transformative change on technological choices, underlining the importance of calibrating energy infrastructures and institutions to serve broader social goals. The agenda of energy transition is closely intertwined with urban climate action, which is equally daunting. Eric Chu and Kavya Michael offer a sobering assessment of climate adaptation action in the urban context, which has been the subject of noted interventions by international donors and multilateral agencies. Yet they show that most donor-supported urban climate programmes conceptualize climate adaptation as a set of top-down technical interventions implemented via public–private mechanisms. Even when such urban climate programmes state that their goal is to address climate vulnerabilities experienced by the most marginalized, the emphasis is on procedural inclusion rather than on addressing the structural factors that shape these unequal exposures.

Arpitha Kodiveri and Rishiraj Sen's examination of India's national and state climate action plans shows that neither the central nor the state governments are alert to the multiple ways in which socioeconomic inequalities relate to India's nascent climate agenda. Their analysis of how national and state climate plans represent concerns of poverty, inequality, gender, and caste-based injustices shows that while several plans recognize gender injustices, few mention caste-related injustices and even fewer mention Dalits. The concerns of poor and marginalized groups are mostly addressed in these plans via the notion of 'co-benefits', which is the assumption that effective climate action will produce ancillary benefits in the form of pollution reduction and easy access to clean energy. However, such assumptions are untenable considering the deeply entrenched caste, class, and gender inequalities that mediate the implementation of all policies and programmes.

Vaishnavi Behl and Prakash Kashwan's contribution offers a snapshot of why both caste and gender inequalities need to be factored into the pursuit of climate justice. They use an intersectional approach to show that the mutually reinforcing effects of gender, caste, and class inequalities determine access to drinking water and opportunities for the further development of drinking water resources in Uttarakhand and Gujarat. Their analysis identifies the specific steps that donor agencies, non-governmental organizations (NGOs), and government agencies can take to address the vulnerabilities faced by Dalits (especially Dalit women) in the midst of climate crisis. Similarly, Ashlesha Khadse and Kavita Srinivasan develop an intersectional approach to study women farmers' collectives in the states of Tamil Nadu and Kerala. They emphasize the importance of an intersectional understanding of how caste, class, age, education, and marital status affect participation in women's land collectives. Further, they offer a comparative analysis of the key differences between the policies and programmes related to women's collective land rights and promotion of agroecological farming pursued by the two state governments. Overall, they find that socioeconomic inequalities and state government policies jointly affect the success of programmes dedicated to climate resilience and agrarian justice.

The omission of caste inequalities is one of the several challenges to India's climate justice movement. Srilata Sircar tackles caste inequalities and injustice head-on by showing that caste hierarchies shape climate vulnerabilities via their effects on land, labour, and spatial relations. She uses examples from the agrarian, urban, and industrial sectors, as well as from India's nascent climate justice activism, which has been largely indifferent toward questions of caste. Building on this extensive engagement, Sircar points to future pathways for reimagining climate justice as caste justice. Prakash Kashwan's analysis of India's three most prominent environmental movements suggests that instead of conceptualizing India's climate or climate justice movement as a monolithic phenomenon, it is important to investigate how diverse, and at times competing, frames and discourses of climate justice shape climate debates in India. For example, one must ask if frequent references to 'co-benefits' as a way to tackle social inequalities in climate plan documents and policy scholarship may have crowded out deeper engagements with questions of equity and justice. This illustrates how climate policy and programmatic choices shape the pursuit of domestic climate justice, a topic that requires deeper investigations.

Policy and programmatic lessons

In this volume, we have highlighted how climate change – in the context of both mitigation and adaptation or resilience-building – calls into question the basic developmental paradigms that underlie policies and plans that India has pursued. The multi-scalar nature of both the climate challenge and potential solutions calls for a coordinated approach that places social equity and justice at the centre of various sectoral policies and programmes. At the national level, there is a need to recalibrate the climate change agenda along its human dimensions, focusing on its implications for housing, infrastructure, ecosystems, food security, health and sanitation, water, education, and economic opportunities. Instead of investing in technology-driven top-down solutions that expose local and state governments to significant debt, key elements of climate action must be developed and implemented via public investments. In addition, these investments must be directed toward building civic capacities and ecological resilience to deal with future changes and uncertainties. In light of the high demand for climate-proof infrastructure, private and non-state financial support may be a necessity in some cases. However, investment decisions should not be based on bankability alone such that they benefit shareholders at the expense of local rural and urban communities. Public–private partnerships must also be designed to deliver long-term social benefits rather than short-term and speculative returns for corporations. Such policies should prioritize inclusive design and collective monitoring of project outcomes. They must focus on empowering historically marginalized groups, including informal workers, residents of informal settlements, women, tribal communities, religious minorities, or the so-called lower castes.

One immediate point of entry for policymakers is to tackle the operational disconnect between climate mitigation and adaptation. Mitigation is often (rightly) prioritized since the global community must first tackle its dependence on fossil fuels, transition to cleaner energy sources, and mobilize collective behaviours to reduce consumption, especially among wealthy groups. However, mitigation's bias towards green technologies and technocratic 'fixes' often lends itself to more financially speculative forms of large-scale infrastructure and investment-led partnerships between private and public sectors. This is especially true in India where governments at all scales seek modern, 'high-tech', and consumption-led forms of economic growth by working in concert with private land developers, transnational corporations, and industrial conglomerates. Emerging critiques of 'smart cities', including Ayona Datta's (2015) work in Dholera, Gujarat; Komali Yenneti and Rosie Day's study on Charanka Solar Park, Gujarat; and Diganta Das's work on Hyderabad HITECH City, highlight how a disposition towards smart technologies and renewable energy can lead to visions of development that are disconnected from the lived experiences of local communities (Datta 2015; Yenneti and Day 2015; D. Das 2015). Instead, this vision speaks to India's desire to be a competitive player in global geopolitics as well as the prioritization of upper-middle-class definitions of environmentalism and quality of life.

Climate adaptation, on the other hand, is more directly linked to poverty alleviation, vulnerability reduction, social empowerment, and community-level access to basic services, which are often deeply enmeshed within social, cultural, political, and economic structures. As a result, climate adaptation priorities tend to be marginalized given the fewer opportunities for them to generate significant financial profits. In India, adaptation priorities continue to play 'second fiddle', especially when compared to the resources, leadership capacity, and scientific expertise dedicated to mitigation efforts. Such a disconnect between mitigation and adaptation leads to the marginalization of the interests and perspectives of frontline communities and their exclusion from decision-making processes. It also confines any potential benefits derived from climate action to those who can afford to invest in or pay into mitigation efforts, while detracting from investments that would protect frontline communities against future climate impacts.

A pivot towards thinking about mitigation–adaptation synergies, equitable resilience, and social transformations is a prerequisite for placing justice at the heart of policymaking. Despite emerging critiques of resilience thinking, there continues to be a push towards pursuing climate-resilient development pathways (CRDPs), as codified in the IPCC *Special Report on Global Warming of 1.5°C* (2018). Conceptually, CRDPs involve a joint development trajectory that both reduces greenhouse gas emissions and builds adaptive capacities to counter ongoing and future climate risks. Practically, it entails social, political, and economic decision-making processes to further sustainable development (IPCC 2018). For many, such an approach seeks to transform dominant development paradigms and actively redress historically entrenched inequalities (Chu et al. 2019; Pelling 2011). It involves actions that tackle systemic and everyday risks experienced by frontline communities (Ziervogel et al. 2017). For India, this means recognizing and contesting drivers of climate injustice within policy decisions and wider discourses in both global negotiations and regional and local governments. Potential strategies include articulating climate policies that are explicitly gender-transformative or anti-discriminatory in terms of class, caste, religious, or tribal identities and pursuing reparative forms of resource and capacity redistribution in the light of historic developmental injustices.

Centring justice in climate policy also requires questioning the primacy of neoliberal financialized growth, especially the kind of jobless economic growth that India has witnessed over the past quarter of a century. Instead of sharing prosperity, such growth exacerbates socioeconomic inequalities and environmental degradation. But this is not questioned in policy debates even though they determine the nature and direction of policymaking. For example, the Government of India's decision to make the Smart Cities Mission the main plank for urban development reflects its faith in high-technology- and high-investment-driven urban development. However, the results of such a strategy have been rather mixed. For instance, Hyderabad's development of a high-tech smart city has produced a 'fragmented metropolitan' where super-premium enclaves with world-class facilities exist as islands alongside the larger metropolitan region, which suffers from a lack of basic civic amenities (D. Das 2015, 57). Similarly, the quest to make Delhi a world-class city is motivated by visions of aesthetic transformation (Bhan 2009). This has led to the eviction and displacement of a million slum residents who were 'declared illegal because they *looked* illegal' (Ghertner 2015, 184, emphasis in the original). Similar patterns of dispossession and exclusion have been reported from smart city developments throughout India. Most noticeably, there are concerns that the modes of neoliberal governance that the smart city approach depends on could significantly undermine the role of democratically elected urban local bodies (Praharaj, Han, and Hawken 2018). This could create challenges for the broader context of policymaking and enforcement in India.

The Indian legal and policy contexts are characterized by a schizophrenic gap. While the Indian Constitution contains progressive environmental and social safeguards, their enforcement remains shockingly poor. The pursuit of neoliberal economic reforms since the early 1990s has greatly exacerbated this implementation gap. In recent years, India's Ministry of Environment, Forests and Climate Change has considerably weakened the Environmental Impact Assessment guidelines, eased the regulatory framework for industrial projects, and sped up the process for granting environmental 'clearances' to mining projects that destabilize ecological systems, increase emissions, and violate the rights of local communities. Such a lackadaisical approach to the enforcement of social and environmental protections can be attributed to a lack of mechanisms to hold the Indian state accountable. These contextual features shape the uptake of climate policies (Kashwan 2015). Accounting for these structural and contextual features of climate policy and action present formidable epistemological challenges for policy researchers, but they have an opportunity to push against conventional approaches of policy research that focus on a narrow set of questions specific to a policy regime.

The multi-scalar nature of the climate change challenge requires expanding policy research to all sectors that stand to be affected by climate impacts and emerging global agreements that note the critical need for sustainable and transformative change. For example, the Sustainable Development Goals (2015), Sendai Framework for Disaster Risk Reduction (2015), UN-Habitat III New Urban Agenda (2016), and the Paris Agreement stocktaking process in 2023 emphasize the need for inclusive and equitable approaches to mitigate or adapt to climate change and build societal resilience. Although these global agreements lack strong enforcement mechanisms, they do articulate broad objectives for the inclusion of women, religious minorities, informal settlers, and indigenous and traditional communities within decision-making. In addition to global agreements, numerous local, regional, and civil society efforts are underway to ensure that climate policies and strategies are equitable and inclusive. In India, NGOs and social movements, such as Slum/Shack Dwellers International, Mahila Housing SEWA Trust, and Mahila Kisan Adhikaar Manch (Forum for Women Farmers' Rights), are increasingly mobilizing for climate justice in the context of housing, women's land rights, and other economic and social rights. Researchers have also documented the emergence of local, community-based efforts that focus on informal settlements, women's groups, the rural poor, Adivasis, Dalits, and other marginalized communities at the frontline of climate impacts (Kothari and Joy 2018). But the intransigence of the state and other powerful actors vested in the status quo hamper the success of these inspiring experiments.

International human rights conventions and declarations highlight the non-negotiability of fundamental rights to life, health, and subsistence, which are threatened by climate change (Caney 2010). This includes the Universal Declaration of Human Rights, adopted by the UN General Assembly in December 1948, which recognizes civil, political, economic, social, and cultural rights. Three recent UN declarations are directly relevant to questions of climate justice in India and elsewhere. First, the United Nations Declaration on the Rights of Indigenous Peoples (UNDRIP), adopted in September 2007, seeks to enshrine the rights that 'constitute the minimum standards for the survival, dignity and well-being of the indigenous peoples of the world'. Second, in July 2010, the United Nations General Assembly passed a resolution to recognize the human right to water and sanitation, calling upon states and international organizations to provide financial resources and assist in capacity-building and technology transfer that some countries in the Global South need for providing safe, clean, accessible, and affordable drinking water and sanitation for all. Third, and most recently, the United Nations Declaration on the Rights of Peasants and Other People Working in Rural Areas (UNDROP), adopted by the United Nations in 2018, offers important levers for regulating projects or activities that affect ecological systems that are central to rural livelihoods (Kashwan, Kukreti, and Ranjan 2021).

None of these declarations can guarantee that states are held accountable, but they put the onus of enforcement on states and powerful market actors, which could provide additional leverage for civil society actors. One important example is the Right to Food campaign spearheaded by India's civil society networks. The campaign drew on the constitutional protection of the right to life and used the judiciary to seek stronger enforcement of right to food provisions (Hertel 2015). But similar mobilizations have not occurred vis-à-vis the rights of internally displaced populations. In 2019 alone, India witnessed over 5 million cases of internal displacement due to natural disasters potentially linked to climate change (*Economic Times* 2020). The UN High Commissioner for Refugees (UNHCR) has asked states and, where relevant, non-state actors to prevent, respond to, and resolve internal displacement while complementing and reinforcing national response efforts. Additionally, the UNHCR report acknowledges the political complexities and challenges presented when displacement is a result of government action or inaction (UNHCR n.d.). Climate policy researchers have an opportunity to investigate how international agreements and declarations can shape domestic policymaking and enforcement – for example, by fostering transnational solidarity networks that seek to hold governments and powerful market actors accountable (Kashwan, Kukreti, and Ranjan 2021). Another important step would be to conceptually ground policy research more strongly in developments in various social science fields, which feature some remarkable research on questions of environmental and climate justice.

Social scientific research agenda

The contributors to this volume highlight the various political dynamics, socioeconomic conditions, and embodied experiences underpinning the struggle for justice in a changing climate. The rich empirical findings and conceptual arguments provide critical analyses of the conditions found in India. However, in a globalized political economy, climate inequality and injustice within the country have widespread implications for the global circulation of knowledge, ideas, resources, and networks. Social scientists in sociology, anthropology, human geography, political science, development studies, urban planning, critical cultural studies, and beyond can, therefore, play an important role in pushing for radical and transformative ideas that are required in the pursuit of climate justice. In this section, we highlight three research frontiers that have the potential to strengthen India's trajectory towards transformative climate justice. These include (*a*) theorizing sources and drivers of climate injustice formation, (*b*) connecting scalar struggles for radical social change, and (*c*) harnessing collective imaginaries of alternative climate futures. We briefly elaborate on these frontiers by distilling the broader theoretical implications of the major findings of this volume.

A theory of climate injustice formation would allow researchers to diagnose the mechanisms through which social exclusion and marginality are created through historical processes and entrenched in contemporary Indian society.

Some policymakers and researchers present the climate change challenge in India as zero-sum, where discursive frames focus on the need to ensure competition, private property rights, economic investment potential, and general entrepreneurial behaviours in governance. In this volume, neoliberalism and the tools and instruments used to promote and entrench it have been critiqued diligently. These important diagnoses show how emerging climate mitigation and adaptation actions are deeply influenced by industrial and financialized development goals. Many social scientists also seek to better theorize the drivers of extractive and speculative sociopolitical practices. These theories offer a deeper reading of India's political–economic history and its contentious relationship with the environment. They also critically interrogate how citizenship and community structures have been shaped by their conflictual relationships with the postcolonial state apparatus. Such ideas point to a need to better understand the fundamental social and political dynamics that underpin the fight for representation, rights, and democratic decision-making in development processes.

Luckily, researchers uncovering the multiple complex ways climate injustice is formed in India can draw inspiration from extensive literatures on rural and agrarian change, political ecologies and geographies of resource extraction, socioeconomic informality, subaltern politics, and alternative and post-development discourses. All of these have a long history of identifying how the concentration of political and economic power has led to the widening of social inequality across India (Roy 2011). Climate injustice, therefore, is a product of India's many developmental inequalities. For instance, there is a need to better consider the impact of different socio-cultural identities and the resultant political and economic disadvantages – especially among historically marginalized groups – when designing and implementing climate solutions. The authors in this volume and beyond have highlighted the need to explore drivers of inequality and processes that entrench them (Michael et al. 2020; Rao et al. 2019). Further, climate injustice should be conceptualized by accounting for multiple, overlapping, and intersectional forms of inequalities among socio-cultural identities and class differences (Cannon and Chu 2021). Such approaches highlight how particular social groups are complexly marginalized and rendered invisible within the state's responses to climate change. The differences in social and political power require institutions and policies that are designed to limit the influence that powerful actors wield within the status quo (Kashwan, MacLean, and García-López 2019).

A second frontier to theorizing climate justice in India is uncovering the knowledge systems, ideas, and practices that help connect and mobilize social struggles across scales (Mehta, Adam, and Srivastava 2022). Researchers of climate change governance have long noted its inherently scalar nature, which makes mobilizing for collective welfare across scales particularly challenging (Revi 2008). India is an example of a country where policy action is formally decentralized – albeit with a strong influence from the central government – and where climate priorities sit at the juncture of multiple sectoral domains, ranging from public infrastructure and health to energy and agriculture (Dubash et al. 2018). Insights from India also reveal how the jurisdictional boundaries of political authorities often do not correspond to the actual spatial expanse of potential mitigation or adaptation actions; as such, decision-making authority pertaining to a cross-sectoral priority like climate change may be devolved across the national, state, and local scales, with no single actor responsible for coordination. From a climate justice perspective, this can lead to gaps in leadership, legal authority, and resource transfer pathways; it can also create opportunities for errant behaviours like political elites exploiting uncertainties and maximizing individual interests on the ground. Historically disadvantaged communities bear the brunt of such forms of exploitation. Yet much of past research does not capture the multiple ways in which social and economic inequalities shape official climate policy debates.

Experiences from India not only highlight how climate priorities can be misaligned but also that the definition of the problem itself can be distorted across national, regional, and local scales, especially given how complex diagnosing the drivers of climate injustice formation can be (Joshi 2014). Struggles for radical social change in the context of climate change must, therefore, bridge the deficits in problem-framing and opportunities to mobilize across scales. Researchers (including those contributing to this volume) have already diagnosed how the central government's push for the 'right to development' in international negotiations is misaligned with the distributive implications of potential climate actions, which can place unequal burdens on historically marginalized and disadvantaged communities (Ziervogel et al. 2017). Going forward, social science research on India's climate policies must focus on identifying the knowledge systems, ideas, and practices that connect citizens to the multi-level state apparatus, focusing on bridging leadership, communication, and capability gaps that inhibit transformational change. This includes developing concrete mechanisms for political intermediation to engage citizens, civil society organizations, and social movements in ongoing policy debates, policymaking, and policy implementation efforts (Kashwan 2017). Such democratization of the policymaking process seems to be a prerequisite for more progressive policies and programmes.

Social science expertise is also crucial for uncovering embodied, cultural, and contextually situated knowledge systems to contest dominant top-down (often engineered) climate solutions, especially ones that support 'green' or 'smart' technological innovation, private investments, and continued growth-oriented strategies. The exploitative and unjust outcomes of top-down climate solutions are well documented; so, to resist them, future research must partner with social movements. Such partnerships are necessary to better track the benefits and losses that result from policy decisions and the method and criteria for these accounting processes. Working with social movements to connect social struggles across scales may help us pivot towards more equitable and inclusive forms of climate-resilient development. Such collaborative efforts to enhance social well-being and empowerment can help redress intergenerational and compounding forms of human vulnerability driven by previously extractive forms of economic growth.

The third frontier is in harnessing and asserting collective imaginaries of alternative climate futures. Scholars of political ecology, environmental sociology, politics, and anthropology are increasingly speaking to alternative development paradigms that move beyond zero-sum thinking in climate action (Gajjar, Singh, and Deshpande 2019). Researchers working to further climate justice in India could explore more radical forms of sustainability transitions and resilience, for example, which can move the focus from financialized growth towards balancing different social and ecological needs (Gerber and Raina 2018). Emerging climate resilience efforts in India are also looking to ecosystem or nature-based solutions, such as coastal mangroves and reforestation projects, to help with carbon sequestration as well as protect communities against extreme hazards such as pluvial flooding, storm surges, and sea-level rise. Not only do nature-based solutions offer mitigation and adaptation co-benefits, but under certain conditions they can also help to empower and regenerate local communities through vocational training and job creation. But past evidence offers reasons to exercise significant caution in the large-scale implementation of nature-based solutions. Critics of resilience thinking have noted how it is only a technical fix that is susceptible to exploitative and exclusionary tendencies (Bahadur and Tanner 2014). There are well-documented cases of displacement of local communities because of land grabs triggered by carbon forestry and government agencies exploiting these programmes to reassert their control in forested areas (Fleischman et al. 2020). Policies to further climate resilience are often captured by local elite interests, which prioritize an economic system that serves the beneficiaries of the status quo, thereby leading to yet another form of greenwashing (Chu 2020). These perverse outcomes of recent sustainability and resilience-building actions in India suggest that developmental pathways should be envisioned more radically, perhaps by directly working with and empowering local communities across rural–urban and class–caste divides, to tackle the root causes of socioeconomic vulnerability and generate alternative visions of the future.

In sum, to gain a deeper understanding of India's climate justice trajectory, social scientific research must pursue advancements in diagnosing the drivers of climate injustice, including its formation and entrenchment, and the role of researchers in informing and/or mobilizing social struggles that bridge scales and offer radically different visions of developmental futures. This will entail working with policymakers, social movements, and historically disadvantaged communities to promote collective social change, redistribute capacities and resources, and redress historic development inequalities. Developing strong research partnerships is therefore critical as we need better theorization of the multiple, overlapping forms of social vulnerability and marginalization in the context of climate change.

Conclusion

The COVID-19 pandemic has shone a spotlight on the sheer grossness of social and economic inequalities in India and their grave consequences for the integrity of public systems at large. As a *Time Magazine* article put it, 'India's vaccine nationalism – along with Prime Minister Narendra Modi's empty showboating – not only plunged India into an unexpected vaccine shortage, but also put countries banking on vaccines from India at great risk' (Roy Chowdhury 2021, para 4). Equally important, India's prowess as the software outsourcing capital of the world, several years of digital governance campaigns, including the roll-out of digital identity cards and e-governance initiatives, have proved to be of little use in battling the pandemic. Researchers have argued that 'technology driven, centralised and surveillance oriented urban regimes' popular among proponents of smart cities have tended to worsen existing inequalities in the face of this unprecedented public health crisis (Gupte et al. 2021, 1). Instead, 'frugal innovations by firms, consumers and city governments' have proved to be far more effective (Gupte et al. 2021, 1). The lesson for India's climate diplomacy and its domestic climate justice could not be clearer.

Persistent debates about international versus domestic climate justice are unhelpful. Instead, scholars, climate activists, and policymakers must investigate and address the complex ways in which international and subnational policies, programmes, and resource mobilizations intersect to influence climate vulnerabilities and the outcomes of specific types of climate policies. In this volume, we have reflected on action pathways pertinent to different sectors of the economy and society in the pursuit of domestic climate justice.

We hope that the research and scholarship agendas we have outlined in this chapter provide helpful pointers to young researchers entering the field at this crucial juncture in Indian and global history. It is important to underline that scholars must seek to pursue publicly engaged research programmes that advance the frontiers of knowledge production while also making significant contributions to the praxis of climate justice. This would require active collaborations with community groups and social movements, whose mobilizations are indispensable for contesting larger development narratives. The aim should be to offer alternative visions of what a climate-changed future *ought* to look like and to provide more radical imaginaries of how, through tackling climate change, we can create a more just and inclusive society for current and future generations. Such efforts, although grounded in local histories, cultures, and social formations, will speak to the larger processes that reinforce societal inequality and poverty in a changing climate. Therefore, the specific insights from learning and theorizing about India might also be fertile grounds for a cosmopolitan reimagination of climate justice theories and movements globally.

References

Adve, N.. 2007. 'Implications of Climate Panel Report'. *Economic and Political Weekly* 42 (12): 1001–1003.

Agarwal, A., and S. Narain. 1991. *Global Warming in an Unequal World: A Case of Environmental Colonialism*. New Delhi: Centre for Science and Environment.

Bahadur, A. and T. Tanner. 2014. 'Transformational Resilience Thinking: Putting People, Power and Politics at the Heart of Urban Climate Resilience'. *Environment and Urbanization* 26 (1): 200–214.

Bhan, G. 2009. '"This Is No Longer the City I Once Knew." Evictions, the Urban Poor and the Right to the City in Millennial Delhi'. *Environment and Urbanization* 21 (1): 127–142.

Bidwai, P. 2012. 'Climate Change, India and the Global Negotiations'. *Social Change* 42 (3): 375–390.

Caney, S. 2010. 'Climate Change, Human Rights, and Moral Thresholds'. In *Climate Ethics: Essential Readings*, edited by S. M. Gardiner, S. Caney, D. Jamieson, H. Shue, and R. K. Pachauri. New York: Oxford University Press, 163–177.

Cannon, C. E. B., and E. K. Chu. 2021. 'Gender, Sexuality, and Feminist Critiques in Energy Research: A Review and Call for Transversal Thinking'. *Energy Research and Social Science* 75. https://doi.org/10.1016/j.erss.2021.102005.

Chu, E. 2020. 'Urban Resilience and the Politics of Development'. In *Climate Urbanism: Towards a Critical Research Agenda*, edited by V. Castan Broto, E. Robin, and A. While. Cham: Springer International Publishing, 117–136.

Chu, E., and K. Michael. 2019. 'Recognition in Urban Climate Justice: Marginality and Exclusion of Migrants in Indian Cities'. *Environment and Urbanization* 31 (1): 139–156.

Chu, E., A. Brown, K. Michael, J. Du, S. Lwasa, and A. Mahendra. 2019. *Unlocking the Potential for Transformative Climate Adaptation in Cities*. Washington, DC, and Rotterdam: World Resources Institute.

Das, D. 2015. 'Hyderabad: Visioning, Restructuring and Making of a High-Tech City'. *Cities* 43: 48–58.

Das, J. 2020. 'The Struggle for Climate Justice: Three Indian News Media Coverage of Climate Change'. *Environmental Communication* 14 (1): 126–140.

Datta, A. 2015. 'New Urban Utopias of Postcolonial India: "Entrepreneurial Urbanization" in Dholera smart city, Gujarat'. *Dialogues in Human Geography* 5 (1): 3–22.

Dubash, N. K., R. Khosla, U. Kelkar, and S. Lele. 2018. 'India and Climate Change: Evolving Ideas and Increasing Policy Engagement'. *Annual Review of Environment and Resources* 43 (1): 395–424.

Fisher, S. 2015. 'The Emerging Geographies of Climate Justice'. *Geographical Journal* 181 (1): 73-82.

Fleischman, F., S. Basant, A. Chhatre, E. A. Coleman, H. W. Fischer, D. Gupta, B. Güneralp, P. Kashwan, D. Khatri, R. Muscarella, J. S. Powers, V. Ramprasad, P. Rana, C. R. Solorzano, and J. W. Veldman. 2020. 'Pitfalls of Tree Planting Show Why We Need People-Centered Natural Climate Solutions'. *BioScience* 70 (11): 947–950.

Gajjar, S. P., C. Singh, and T. Deshpande. 2019. 'Tracing Back to Move Ahead: A Review of Development Pathways that Constrain Adaptation Futures'. *Climate and Development* 11 (3): 223–237.

Gerber, J. F., and R. S. Raina. 2018. *Post-Growth Thinking in India: Towards Sustainable Egalitarian Alternatives*. Hyderabad: Orient Blackswan.

Ghertner, D. Asher. 2015. 'Why Gentrification Theory Fails in "Much of the World"'. *City* 19 (4): 552–563.

Gupta, J. 2014. *The History of Global Climate Governance*. Cambridge: Cambridge University Press.

Gupte, J., S. M. G. Babu, D. Ghosh, E. Kasper, P. Mehra, and A. Raza. 2021. *Smart Cities and COVID-19: Implications for Data Ecosystems from Lessons Learned in India*. Brighton: Social Science in Humanitarian Action (SSHAP).

Hertel, S. 2015. 'Hungry for Justice: Social Mobilization on the Right to Food in India'. *Development and Change* 46 (1): 72–94.

IPCC. 2018. *Global Warming of 1.5 °C: An IPCC Special Report on the Impacts of Global Warming of 1.5 °C above Pre-industrial Levels and Related Global Greenhouse Gas Emission Pathways, in the Context of Strengthening the Global Response to the Threat of Climate Change*. Geneva, Switzerland: Intergovernmental Panel on Climate Change (IPCC).

Jasanoff, S. 1993. 'India at the Crossroads in Global Environmental Policy'. *Global Environmental Change* 3 (1): 32–52. doi: https://doi.org/10.1016/0959-3780(93)90013-B.

Joshi, S. 2014. 'Environmental Justice Discourses in Indian Climate Politics'. *GeoJournal* 79 (6): 677–691.

Kashwan, P. 2015. 'Forest Policy, Institutions, and REDD+ in India, Tanzania, and Mexico'. *Global Environmental Politics* 15 (3): 95–117.

———. 2017. *Democracy in the Woods: Environmental Conservation and Social Justice in India, Tanzania, and Mexico*. New York: Oxford University Press.

Kashwan, P., L. M. MacLean, and G. A. García-López. 2019. 'Rethinking Power and Institutions in the Shadows of Neoliberalism (An Introduction to a Special Issue of World Development)'. *World Development* 120: 133–146.

Kashwan, P., I. Kukreti, and R. Ranjan. 2021. 'The UN Declaration on the Rights of Peasants, National Policies, and Forestland Rights of India's Adivasis'. *International Journal of Human Rights* 25 (7): 1184–1209.

Kothari, A. and K. J. Joy 2018. *Alternative Futures: India Unshackled*: New Delhi: Authorsupfront Publishing Services Private Limited.

Mehta, L., H. N. Adam, and S. Srivastava. 2022. *The Politics of Climate Change and Uncertainty in India*. London and New York: Taylor & Francis.

Michael, K., M. K. Shrivastava, A. Hakhu, and K. Bajaj. 2020. 'A Two-Step Approach to Integrating Gender Justice into Mitigation Policy: Examples from India'. *Climate Policy* 20 (7): 800–814.

Praharaj, S., J. H. Han, and S. Hawken. 2018. 'Urban Innovation through Policy Integration: Critical Perspectives from 100 Smart Cities Mission in India'. *City, Culture and Society* 12: 35-43. https://doi.org/10.1016/j.ccs.2017.06.004.

Pelling, M. 2011. *Adaptation to Climate Change: From Resilience to Transformation*. London and New York: Routledge.

PTI. 2020. 'Over 5 Million People Internally Displaced in India in 2019: UN'. *Economic Times*, 5 May. https://economictimes.indiatimes.com/news/politics-and-nation/over-5-million-people-internally-displaced-in-india-in-2019-un/articleshow/75548906.cms (accessed 24 December 2021).

Rao, N., A. Mishra, A. Prakash, C. Singh, A. Qaisrani, P. Poonacha, et al. 2019. 'A Qualitative Comparative Analysis of Women's Agency and Adaptive Capacity in Climate Change Hotspots in Asia and Africa'. *Nature Climate Change* 9 (12): 964–971.

Revi, A. 2008. Climate Change Risk: An Adaptation and Mitigation Agenda for Indian Cities'. *Environment and Urbanization* 20 (1): 207–229.

Roy, A. 2011. 'Slumdog Cities: Rethinking Subaltern Urbanism'. *International Journal of Urban and Regional Research* 35 (2): 223–238.

Roy Chowdhury, D. 2021. 'Modi Never Bought Enough COVID-19 Vaccines for India. Now the Whole World Is Paying'. *Time*, 28 May. https://time.com/6052370/modi-didnt-buy-enough-covid-19-vaccine/ (accessed 24 December 2021).

Shue, Henry. 1993. 'Subsistence Emissions and Luxury Emissions'. *Law and Policy* 15 (1): 39–60.

UNHCR. 'Internally Displaced People'. https://www.unhcr.org/en-us/internally-displaced-people.html (accessed 24 December 2021).

Yenneti, K. and R. Day. 2015. 'Procedural (In)justice in the Implementation of Solar Energy: The case of Charanaka Solar Park, Gujarat, India'. *Energy Policy* 86: 664–673. https://doi.org/10.1016/j.enpol.2015.08.019

Ziervogel, G., M. Pelling, A. Cartwright, E. Chu, T. Deshpande, L. Harris, et al. 2017. 'Inserting Rights and Justice into Urban Resilience: A Focus on Everyday Risk'. *Environment and Urbanization* 29 (1): 123–138.

About the Editor and Contributors

Editor

Prakash Kashwan is Associate Professor of Environmental Studies at Brandeis University, Waltham, Massachusetts. At the time of preparation of this volume, he was Associate Professor in the Department of Political Science and Co-Director of the Research Program on Economic and Social Rights, Human Rights Institute, University of Connecticut, Storrs. His research and scholarship build on nearly a quarter century-long engagement with questions at the intersection of environmental and social justice, including via a first career in international development (1999–2005). He is the author of *Democracy in the Woods: Environmental Conservation and Social Justice in India, Tanzania, and Mexico* (2017), editor of the journal *Environmental Politics*, and a member of the editorial boards of *Progress in Development Studies, Earth System Governance, Sage Open,* and *Humanities and Social Sciences Communications.* He has also contributed popular commentaries to the *Washington Post, The Guardian, The Conversation, Al Jazeera, Africa Is a Country, The Wire, Down to Earth,* and the *Hindustan Times.*

Contributors

Vaishnavi Behl is a Mumbai-based writer and researcher studying access to food and water to marginalized communities in India. She has an interdisciplinary background having studied sociology and anthropology at St. Xavier's College, Mumbai, and international relations at Sciences Po, Paris. Growing up in the midst of deep economic and social disparities in Mumbai, she has always aspired to bring to light issues surrounding unequal access to basic facilities to a wider audience. While adding to existing knowledge on inequalities in India, she aims to combine grounded research with community-level interventions. She is passionate about using her skill set to make academic research engaging and accessible to a wider audience, with a goal of bringing about lasting and meaningful change.

Parth Bhatia is an Associate Fellow at the Centre for Policy Research, New Delhi, working with the Initiative on Climate Energy and Environment (ICEE). His interests include the political economy of India's electricity sector at the state level, national socio-technical energy transitions and the design of effective climate institutions and policies. He is a contributing author for the chapter on 'National and Sub-national Climate Policies and Institutions' in the Working Group 3 report of the IPCC AR6 cycle. He holds M.Tech and B.Tech. Degrees in Energy Systems Engineering from the Indian Institute of Technology Bombay. Prior to joining CPR, Parth worked as a Business Analyst at A.T. Kearney India.

Haimanti Bhattacharya is an Associate Professor of Economics at the University of Utah. She received her doctorate in Economics from the University of Arizona and has been a post-doctoral fellow at the Earth Institute at Columbia University. She is an applied micro-economist with research interests in environment, resource and food, behavioral economics, and gendered well-being. She has been a recipient of the Quality of Research Discovery Award awarded by the European Association of Agricultural Economists. A common underlying thread that ties the wide range of her research topics is the challenges for the sustainability of human society – economic, social, and environmental sustainability. India is the primary geographic focus of her research.

Vasudha Chhotray is Professor of Politics and Development in the School of International Development at the University of East Anglia, Norwich, UK. She is a political scientist interested in the politics and governance of natural resource use and extraction, and the larger politics of development, welfare, and citizenship in India. In addition to the political economy of green energy, just energy transitions and the future of coal in India, her current research interests extend to the long-term trajectories of disaster recovery in India. She has published in leading academic journals, most recently in *World Development, Development and Change, Geoforum, Society and Natural Resources,* and *Contemporary South Asia.* Her monograph *The Anti-politics Machine in India* was published in 2011. She regularly contributes opinion pieces drawing from her research to critical media like the *Conversation UK* and *Scroll.in.*

Eric Chu is an Assistant Professor in the Department of Human Ecology and Co-Director of the Climate Adaptation Research Center at the University of California, Davis. He researches how cities design policies to address climate change. He has written extensively on urban climate resilience, particularly on issues of socio-spatial change, development planning, governance experimentation, and environmental justice. He is also a Lead Author in Working Group II of the Sixth Assessment Report to the Intergovernmental Panel on Climate Change (IPCC) and Chapter Lead for the Fifth US

National Climate Assessment (NCA5). He holds a PhD in Environmental Policy and Planning from the Massachusetts Institute of Technology.

Ashlesha Khadse is part of the Amrita Bhoomi Centre, a farmer-led agroecology school in India affiliated to the Karnataka State Farmers Movement (KRRS for its initials in Kannada language) and La Via Campesina, the global network of farmers movements. She completed her Master of Science in Agroecology at El Colegio de la Frontera Sur (ECOSUR) in Chiapas, Mexico, where she conducted research on how peasant movements scale up agroecology. She has published research papers on agroecology movements and policies in India. Ashlesha has been associated with farmers' movements in India since 2009.

Arpitha Kodiveri is a Hans Kelsen Fellow and has a PhD from the European University Institute. Her research work examines land conflicts and legal mobilization by forest-dwelling communities in the extractive economy of Odisha. She has an LLM in European and comparative law from the EUI and an LLM in environmental law from U.C. Berkeley School of Law as a Fulbright–Nehru Fellow. She has previously worked as a senior research associate at the Ashoka Trust for Research in Ecology and the Environment and as an environmental lawyer with Natural Justice on forest rights and governance questions in India.

Kavya Michael is an environmental social scientist affiliated as a postdoc researcher at Chalmers Technical University, Sweden. Her expertise lies in examining climate change through a human rights and justice lens. She researches on the multiple intersections of climate change/environmental hazards, informality, gender, urban inequality, inter and intra-regional migration, and development in India. She is a recipient of the IDRC Climate Leadership Fellowship and was hosted in the Political Economy Research Institute at the University of Massachusetts, Amherst. Recently she has engaged significantly with the question of gendered injustices in the climate change space as well as the need to design a transformative transition, to gender just low carbon economic pathways.

Rishiraj Sen is a final year undergraduate student at Guwahati College (affiliated to Gauhati University), currently pursuing his bachelor's in English Literature. He has published three poetry books under the pen name Sutputra Radheye. He has written on cultural and political issues for several publications, such as *Frontier*, *The Quint*, and *Countercurrents*. His research interests are Marxism and society, postcolonialism, media, climate justice, and democracy, among other topics.

Karnamadakala Rahul Sharma is an Associate Fellow with the State Capacity Initiative at the Centre for Policy Research in New Delhi and a PhD candidate at the Bren School of Environmental Science and Management at the University of California, Santa Barbara. He works on administrative reform, organizational norms and culture in the bureaucracy, and the role of non-financial incentives in motivating pro-social and pro-environmental behaviour. Rahul has worked for a decade across engineering, monitoring and evaluation, and public policy research with a sectoral focus on energy access, climate change, and sustainable cities. He has published widely on these topics, including in the journals *Science*, *Nature Energy*, and *Energy for Sustainable Development*. He is an Associate Editor for the *Journal of Environment and Development* and Lead Author of UNEP's Global Environmental Outlook on Cities 2021.

Srilata Sircar is a Lecturer in India and Global Affairs at the King's India Institute, King's College London. She has previously held research positions at Lund University, Sweden, and University College London, UK. Her research interests lie at the intersection of urban political ecology, feminist geography, and postcolonial critique. She currently leads a research group on Cities, Climate, and Capital in the Indian Ocean World. She is the convener of the Confronting Caste series, which is a set of conversations in podcast and panel formats that seeks to centre the analytic of caste in historical and contemporary contexts. Her work has been published in leading journals such as *Geoforum* and *Gender, Place and Culture*, and media portals such as *The Conversation* and *Feminism in India*.

Kavitha Srinivasan is a member of the Mahila Kisan Adhikaar Manch (MAKAAM), which is a network for advancing women farmers rights in India. She has carried out several social audits for multiple government policies in India such as the Mahatma Gandhi National Rural Employment Guarantee Act (MNREGA), mid-day meal schemes, and land inventory in Andhra Pradesh. She is also currently an assistant researcher on 'Decarbonising Electricity: A Comparison in Socio-ecological Relations' with the University of Technology in Sydney.

About the Poets and Artists

Anupriya is an artist from Bihar who lives in Faridabad. She has created artwork for about 70 book covers for renowned publication houses. Anupriya's work is focused mainly on women's issues.

Veena Chhotray worked for the Indian Administrative Service for four decades. She now devotes herself to literary translation.

Nidhin Donald is a Postdoctoral Fellow with the Department of Humanities and Social Sciences at the Indian Institute of Technology (IIT) Bombay. His scholarly interests revolve around studies of family, household, religion, and decennial censuses.

Praneeta Mudaliar is an Assistant Professor of Environmental Studies and Sciences at Ithaca College, New York. Her research and teaching is centered around power dynamics, climate justice, and decolonizing conservation. Praneeta enjoys writing poems, journaling, and long walks.

Late Gail Omvedt was an American-born Indian sociologist and human rights activist. She was a prolific writer and scholar-activist who stood in support of grassroots anti-caste movements and working women's struggles in India.

Bharat Patankar is a leading activist, co-founder, and president of Shramik Mukti Dal, who has worked for nearly 40 years in movements of workers, farmers, dam evictees, agricultural labourers, women's liberation, rights of farmers in windmill development, and radical anti-caste movements.

Samvedna Rawat is a Mumbai-based poet, writer, and storyteller. Samvedna's first book, a collection of poems entitled लड़कियाँ जो पुल होती हैं (lit., Girls Who Become Bridges) was published in 2007 and was the winner of Madhya Pradesh's prestigious Wagheshwari award.

Waharu Sonavane is an eminent Adivasi poet, activist, a member of Shramik Mukti Dal (Toilers' Liberation League), and the co-founder of Adivasi Ekta Parishad (Adivasi Unity Council). गोधड (lit., Quilt), a collection of poetry by Sonavane, was published in 2006 by Pune-based Sugawa Publishers.

Index